Lecture Notes in Economics and Mathematical Systems 597

Christiane Barz

Risk-Averse Capacity Control in Revenue Management

With 32 Figures and 10 Tables

Springer

Christiane Barz
Universität Karlsruhe (TH)
Fakultät für Wirtschaftswissenschaften
Kaiserstraße 12
76128 Karlsruhe
Germany
barz@wior.uni-karlsruhe.de

Dissertation, genehmigt von der Fakultät für Wirtschaftswissenschaften der Universität Fridericiana zu Karlsruhe, gefördert durch die DFG. Referent: Prof. Dr. Karl-Heinz Waldmann; Korreferentin: Prof. Dr. Marliese Uhrig-Homburg; Tag der mündlichen Prüfung: 19.12.2006.

Library of Congress Control Number: 2007930764

ISSN 0075-8442

ISBN 978-3-540-73013-2 Springer Berlin Heidelberg New York

Springer is a part of Springer Science+Business Media

springer.com

Production: LE-TEX Jelonek, Schmidt & Vöckler GbR, Leipzig
Cover-design: WMX Design GmbH, Heidelberg

SPIN 12075102 88/3180YL - 5 4 3 2 1 0 Printed on acid-free paper

Acknowledgements

"A fool . . . is a man who never tried an experiment in his life."
(Erasmus Darwin)

I would like to thank all of those who supported me during my biggest experiment so far, the completion of this thesis.

In particular, I would like to thank my supervisor, Prof. Dr. Karl-Heinz Waldmann. Without his help, this work would not have been possible. I also thank my co-advisor, Prof. Dr. Marliese Uhrig-Homburg, for her valuable and constructive comments on my work and beyond.

I am deeply indebted to Dr. Alfred Müller. His stimulating suggestions and unflinching encouragement were invaluable. I thank him and all the other colleagues at the chair for their support.

This work was funded by the DFG Graduate School for Information Management and Market Engineering. Many thanks go to all members of this program.

Words fail me to express my appreciation to Thorsten Friedrich for his patience, love, and persistent confidence in me.

Finally, I thank my parents and my brother for their faith and constant support.

Christiane Barz Karlsruhe, June 2007

Contents

List of Figures

List of Tables

1 Introduction

"If necessity is the mother of invention, then deregulation is the father, and revenue management (also known as yield management) is the couple's golden child – at least as far as operations research is concerned." (Horner, 2000, p. 47)

Deregulation had a significant impact on the U.S. airline industry in the late 1970s. Charter and low-cost airlines such as People Express and Southwest were able to offer seats at a fraction of the price charged by established carriers like Pan Am and American Airlines. Due to their different cost structure, it seemed to be impossible for the big carriers to offer tickets at the same low price. Yet they had to find a way to compete.

Robert L. Crandall from American Airlines is widely credited with the solution to the problem: *yield management* – today called *revenue management*, since it maximizes revenue earned on a flight rather than yield (revenue per passenger mile).

The idea was simple: American Airlines flights were only half full on average. Offering the empty seats at a discount price would not only enable the carriers to compete with the low-cost airlines but even create additional revenue, if (1) it were possible to prevent cannibalization, i.e. the sale of discount tickets to consumers who would otherwise be willing to pay full fare, and if (2) it could be assured that only the seats that would otherwise fly empty were sold at the low price.

Implementing this strategy, American Airlines matched the low-cost airlines' prices with a limited number of seats that had to be booked several weeks or months in advance. Due to this purchase restriction – the lack of flexibility – the offer was not attractive for the late-arriving demand (typically business travelers) willing to pay the full fare.

Note, however, that if too many seats were sold at low prices, the airline would run the risk of filling the plane too early and losing full-fare customers (risk of revenue dilution). On the other hand, they were taking the chance that discount demand would be rejected, with full-fare demand not sufficing to fill the airplane. The plane would then depart with more empty seats than necessary (demand spoilage).

The ability of American Airlines to control the availability of discount seats had a dramatic effect on its low-cost competitors. People Express was hit especially hard. For details on the "battle" of American Airlines versus People

Express, see Cross (1998, Chap. 4). Revenue management contributed not only to the bankruptcy of People Express but to the demise of several other carriers as well. At the same time, it generated significant additional revenue for the airlines that applied it. According to Smith et al. (1992), American Airlines estimates that over a three-year period at the end of the 1980s, quantifiable benefits of over 1.4 billion dollars were attributable to the control of discount price capacity and overbooking (i.e. selling more reservations than there are seats on the plane). Today, revenue management is both prevalent and mature in the airline industry. In fact, Talluri and van Ryzin (2004b, p. 10) state that the additional revenue generated by revenue management practices accounts for 4 to 5 percent of overall revenue, a value roughly comparable to many airlines' profits in a good year.

One major factor that enabled American Airlines to effectively apply revenue management practices was the use of information technology, namely central reservation systems, to manage the sale of seats. In addition to recording the number of seats sold and the number left to sell, central reservation systems also enabled better price and inventory management.

Capacity control mechanisms allow airlines to open and close the offer of discount fares depending on the number of seats still available, the time remaining until departure, and demand forecasts. Usually, these mechanisms are deeply embedded in the software logic and are expensive and difficult to change (Talluri and van Ryzin, 2004b, p. 28). According to Zhang and Cooper (2005) nested protection levels dominate airline practice due to the fact that many distribution channels allow only these types of controls.

A protection level y specifies the number of seats to reserve (protect) for a particular class or set of classes. If the plane's capacity was 100 and the protection level for full-fare demand was 70, a maximum of 30 seats could be sold at a discount price. Beyond this limit, the discount fare class is closed. In this example, "nested" means that full-fare demand has access to all the capacity reserved for lower fare demand. So if e.g. only 10 seats were sold at the discount fare but there is a high demand for seats at the full price, the airline could sell up to 90 seats at the full fare even if some of them had originally been assigned to the discount fare class. In the case of partitioned (non-nested) classes, only 70 seats would be offered at the full fare. As nested protection levels are so common in practice, one could argue that the formulation of a capacity control problem should require the selection of a control by protection levels (Zhang and Cooper, 2005). Yet the question remains whether (and if, when) this is restrictive.

These central reservation systems constitute one of the earliest examples of e-commerce. Based on the airline industry's success story, it is expected that the use of revenue management will be enhanced by the emerging role of Internet-based e-commerce, see Copeland and McKenney (1988), Smith et al. (2001), Baker et al. (2001), Boyd and Bilegan (2003), and Klein and Loebbecke (2003). But also apart from Internet-based e-commerce, today's information technology is enabling more and more industries to adopt revenue management practices. The hotel and hospitality sector, cruise lines, event promotion firms, and car rental companies are all examples of traditional applications that could be modeled in a similar way to the airline industry. Today, however, entities as diverse as the broadcasting, hospital, casino, and utility industries are starting to use revenue management practices. For surveys on the variety of applications, see McGill and van Ryzin (1999), Talluri and van Ryzin (2004b, Chap. 10), Yeoman and McMahon-Beattie (2004), Kimms and Klein (2005), and Sfodera (2006).

In the recent literature, the term "revenue management" encompasses more general demand management decisions, covering not only the determination of the number of tickets to offer at low prices (which is then called *capacity control*, *seat inventory control*, or *discount allocation*) but also *overbooking*. In addition to these quantity decisions, some authors even extend the meaning of revenue management to include the problem of product creation to ensure that full-fare demand is not sacrificed by offering tickets at discount prices (*market segmentation*) as well as the problem of finding the right prices and adjusting them over time (*dynamic pricing*).

For a survey of the different subtopics, see Kimes (1989), Weatherford and Bodily (1992), Harris and Peacock (1995), Weatherford (1998), McGill and van Ryzin (1999), Boyd and Bilegan (2003), Talluri and van Ryzin (2004b) and Phillips (2005). Different approaches to quantity decisions are summarized in Zehle (1991), Daudel and Vialle (1992), Klein (2001), and Tscheulin and Lindenmeier (2003). Dynamic pricing surveys can be found in Chan et al. (2004), Elmaghraby and Keskinocak (2003), and Bitran and Caldentey (2003); for legal aspects, see Weiss and Mehrotra (2001). Hybrid approaches that deal with both capacity allocation and optimal pricing of the fare classes can be found in Weatherford (1997) and Feng and Xiao (2006b). Badinelli (2000), Walczak (2001), Chatwin (2002), Chatwin (2003), and Maglaras and Meissner (2006) discuss the differences and similarities between capacity control and dynamic pricing. Note that dynamic pricing problems in revenue management do not consider replenishment; surveys on dynamic pricing and inventory decisions are provided by Elmaghraby and Keskinocak (2003) and Chan et al. (2004). Issues of price discrimination in the context of revenue management are discussed e.g. in Faßnacht and Homburg (1998); Talluri and van Ryzin (2004b, Chap. 8) furnish an overview.

To prevent misunderstanding, we will call the above-mentioned practice of determining the number of seats to protect for full-fare demand the "capacity control problem in revenue management", or "capacity control" for short. Overviews that exclusively deal with mathematical models for capacity control can be found in Ben-Yosef (2005, Chap. 7), Pak and Piersma (2002), and Kimms and Müller-Bungart (2004).

A variety of problems is summarized under the term revenue management, and its techniques are applied in many different industries. This work aims neither at an industry-specific nor at an all-embracing approach to revenue management. Instead, the goal is to provide a deeper understanding of the generic single-resource capacity control problem that forms the basis of many revenue management systems. The general terminology of capacity control will be introduced, but for simplicity, we will stick to the terminology of the airline industry thereafter.

1.1 The Basic Capacity Control Problem

The basic capacity control model is concerned with making efficient use of a certain, fixed capacity C of a single resource with homogeneous units that becomes worthless after a given time T.

The company sells its capacity as $i_{\max}$ distinct products. Each product $i = 1, \ldots, i_{\max}$ consists of one unit of this resource and is offered at a price (or fare) of ϱ_i. Without loss of generality, we assume that the products are indexed such that $0 < \varrho_{i_{\max}} \leq \cdots \leq \varrho_i \leq \varrho_1$.

In the airline industry, the capacity could be the number of seats in the economy compartment of a single-leg flight, i.e. a non-stop flight from one origin to one destination departing at some future point in time. All products, also called booking or fare classes, represent one (reservation for a) seat on that flight. They might only differ in price and/or in purchase restrictions such as a Saturday night restriction or early booking conditions. These artificial differences, known as "fencing conditions", ensure that the same resource can be sold at different prices. In the above-mentioned simplified case of American Airlines, there were $i_{\max} = 2$ products or booking classes: tickets at a discount fare that were available only several weeks before departure and the full-fare tickets. We assume the fencing conditions as well as the product prices to be fixed and exogenously given.

The question of capacity control is which products to offer for sale at a given point in time. Frequently, it is advantageous (and feasible) to reword this question and decide whether one should accept or reject an incoming request for product i given a certain amount of remaining capacity and time until departure.

1.1.1 Assumptions

We speak of a basic capacity control problem, if the following assumptions are made:

i) After a certain time T, the whole amount of capacity C is worthless. No additional units of capacity can be ordered.
ii) It is assumed that the major part of the costs is already sunk and that variable costs are negligible, so that the aim of profit maximization can be approximated by maximizing the revenue gained from the selling process.
iii) The resource has to be allocated dynamically as demand materializes. Rejected demand is lost and cannot be stored for the future. Once accepted, a customer cannot be rejected later without significant cost.

iv) There is considerable uncertainty about the quantity and the type of future demand. Future demand for the products offered can be described in terms of a random variable with a known probability distribution.
v) Products consist of a single (homogeneous) resource. Products that are composed of multiple resources, called network problems, are not considered.
vi) The company has monopolistic market power and customers are myopic.
vii) Demand (i.e. the number of requests) and time are discrete.
viii) Group bookings that have to be completely accepted or rejected are not considered. If there is demand for more than one ticket at a time, this demand may be partially accepted.
ix) Customers do not cancel (strictly) prior to the time of service. No-shows, i.e. customers that do not show up at the time of service, are not considered.
x) Demand for the products is independent of the availability of other products.
xi) The decision-maker's preferences can be approximated by a maximization of expected revenue; he is assumed to be risk-neutral.

This basic single-resource capacity control model does not reflect the state of the art in the revenue management literature, but it forms the basis for a lot of more advanced models and for most models used in practice.

Assumptions i), ii), and iii) form the heart of the capacity control problem in revenue management: a fixed amount of perishable capacity, high fixed costs and non-storable demand. (If it were possible to store demand, one would store all of it and sell seats right before capacity perished in decreasing price order.) If variable costs cannot be neglected, the product's contribution margin can usually be considered instead of the price (Zehle, 1991).

In some recent articles, however, the development of new products is propagated to allow the seller to reject some of the demand that has already been accepted against a certain compensation or to reassign it to a different type of capacity; see Biyalogorsky et al. (1999), Biyalogorsky and Gerstner (2004), Gallego and Phillips (2004), and Gallego et al. (2004). The latter products are frequently referred to as "flexible products" and are actually used in the hotel and cruise line industries. Generally, however, concerns about public image preclude their application (Biyalogorsky et al., 1999).

Although forecasting is considered an important ingredient for successful implementation, it is usually omitted from capacity control models. Yet assumption iv) is critical if new routes are offered or schedules are changed. The problem is discussed in van Ryzin and McGill (2000) and solved by an adaptive algorithm. To the author's knowledge, Bayesian demand learning is only considered in the dynamic pricing context; see e.g. Farias and Van Roy (2006) and the references given there. For a different perspective on capacity control that can do completely without demand forecasts, see Ball and Queyranne (2006).

Assumption v) is even more crucial in practice. Many large carriers operate on a flight network with hubs. Due to the curse of dimensionality, network capacity control is often tackled with approximate dynamic programming, heuristics, or simulations. For the basics on capacity control for flight networks, see Phillips (2005, Chap. 8) or Talluri and van Ryzin (2004b, Chap. 3) and the references given there. When the existence of more compartments (such as the business and economy compartments) is considered explicitly, issues of upgrading are discussed e.g. in Biyalogorsky et al. (2005) and Lukaschewitsch (2005). Although revenues are reported to improve by an additional 2.5 percent when network carriers optimize on the network level rather than on single-legs, single-leg models are still widely used (Talluri and van Ryzin, 2004b, p. 82). In addition, they form important building blocks in many heuristics for the network case (Talluri and van Ryzin, 2004b, p. 27).

Assumption vi) is standard in revenue management models. The first model to consider a basic capacity control model in a competitive framework is Netessine and Shumsky (2005). Strategic customers are considered in Anderson and Wilson (2003) and Liu and van Ryzin (2005).

To facilitate mathematics, some models assume demand to be continuous in contrast to the first part of assumption vii), see e.g. Curry (1990), Belobaba (1987a), or Bodily and Weatherford (1995). However, discrete demand seems more natural in real applications. The assumption of discrete time is not a hard to implement, since time can be discretized e.g. by counting arbitrarily small time intervals, by uniformization (Lippman, 1975), or by looking only at the times when demand materializes (as Lin, 2004, did in a dynamic pricing model). The latter approach turns the planning horizon into a random variable representing the number of points in time when demand arrives. Although most capacity control models use a discrete time approach, Liang (1999), Zhao and Zheng (2001), and Feng and Xiao (2006a) use a continuous time approach; a semi-Markov decision process is modeled in Walczak (2001) and Brumelle and Walczak (2003).

Papastavrou et al. (1996), Kleywegt and Papastavrou (1998, 2001) and van Slyke and Young (2000) demonstrate that the capacity control problem can also be formulated as a (stochastic) knapsack problem. They use this approach to handle group bookings that must be accepted or rejected as a whole. Lee and Hersh (1993) and Brumelle and Walczak (2003) consider group bookings within the framework of Markov decision processes. Among other things, they show that capacity control mechanisms that are suitable under assumption viii), such as the control by protection levels, are not optimal for total accept/deny decisions. According to Farley (2003, p. 155), small group bookings are usually either treated as individual bookings; airlines rely on manual processes to price and book larger groups. Eguchi and Belobaba (2004) give a recent overview of the literature on implemented group booking processes in airline revenue management.

In contrast to assumption ix), a significant proportion of tickets are canceled (strictly) prior to departure in airline applications. In addition, some passengers are no-shows, i.e. they simply do not show up at the time of departure. Airlines usually respond to this issue by overbooking: They sell more tickets than they have seats. This is why in the case of cancelations and no-shows, one is strictly speaking of selling reservations or tickets instead of seats. Without overbooking, "about 15 percent of seats would be unused on flights sold out at departure" (Smith et al., 1992, p. 9). While overbooking generates additional revenue, it introduces the risk that more passengers may show up for a flight than there are seats on the plane. Denied boardings, i.e. passengers not boarded because seats are not available, must be compensated monetarily (see e.g. EU-Regulation No 261/2004) and cause customer dissatisfaction (Lindenmeier and Tscheulin, 2005). Talluri and van Ryzin (2004b, Chap. 4) discuss the operational as well as legal issues of overbooking and provide a comprehensive literature survey. For a historical perspective on overbooking, see Rothstein (1985).

Given a single resource problem, there are basically two ways of modeling cancelations and no-shows. On the one hand, one can disentangle the overbooking problem from the capacity control problem, determine an adequate number of tickets $\bar{C}$ (the virtual capacity) to sell, and solve the capacity control problem by pretending that one has a capacity of $\bar{C}$. Basic capacity control models can be used after the virtual capacity has been determined. On the other hand, one could, of course, directly model cancelations and no-shows together with the capacity control problem. See Subramanian et al. (1999), Walczak (2001), and Feng and Xiao (2006a) for such a combined approach.

Assumption x) is named the "independent demand assumption" and is used in most capacity control models. It implies that fencing conditions are chosen such that the products $i = 1, \dots, i_{\max}$ segment the market perfectly into $i_{\max} + 1$ segments $0, \dots, i_{\max}$ – one segment for each booking class and one segment for people not buying any of these $i_{\max}$ products, labeled segment 0. Customers in each segment i buy only the product corresponding to this segment, while customers in segment 0 never buy any product $i = 1, \dots, i_{\max}$. If the corresponding product is not offered at the time of a customer request, the customer does not buy at all and is considered lost.

This approach was initially said to work well in practice. In recent years, however, more and more evidence has been reported against it. Customers are observed to have become increasingly price-sensitive and tend to choose the cheapest option available. The increasing use of the Internet as a distribution channel strengthens this effect, since prices can be easily compared. Hence, an adequately high level of market segmentation is more difficult (if not impossible) to achieve nowadays; see Boyd (2004) and Boyd and Kallesen (2004). The potential pitfalls of ignoring the assumption of perfect market segmentation are modeled in Cooper et al. (2006). The first capacity control models to provide an analytical account of the effect of customer choice between fare classes in a two-class model (in which customers arrive in a strict low-to-high fare order) are described in Pfeifer (1989), McGill (1989), and Brumelle et al. (1990). For a a heuristic extension to more fare classes, see

Belobaba and Weatherford (1996). Zhao and Zheng (2001) analyze a two-class model in which customers are allowed to arrive in any order; for a multi-class model, see Talluri and van Ryzin (2004a). Other models that incorporate simple choice models are Bodily and Weatherford (1995), Botimer and Belobaba (1999), and Corsten and Gössinger (2005). Customer choice between parallel flights (but not between fare classes) is analyzed in Zhang and Cooper (2005).

In this work, we put a special emphasis on assumption xi), which postulates that only the expectation is relevant. Other factors, such as load factor (i.e. the fraction of seats sold at departure) or passenger spill (i.e. the fraction of full-fare requests that must be rejected), need not be considered.

These factors are frequently incorporated by means of an adequate manipulation of the fare prices; see e.g. Brumelle et al. (1990). But even in the latest literature on capacity control expected revenue is the most widespread optimality criterion in use, i.e. a risk-neutral decision-maker is modeled and discounting is not considered. The reason for maximizing undiscounted values is that the planning horizon is usually only a few months. Some authors, such as Subramanian et al. (1999), note that the analysis with discounting is essentially the same, but do not consider discounting explicitly. Risk-neutrality, however, is a harder assumption. When it comes to risk-neutral decision-makers, many capacity control models have been studied extensively, structural properties of the optimal control policy have been proven, heuristics have been promoted, and various extensions and alternatives have been suggested. Yet a basic capacity control model with a risk-averse decision-maker has only been tackled heuristically in Weatherford (2004); a worst-case relative regret criterion was used in Ball and Queyranne (2006).

1.1.2 The Attitude Towards Risk in Capacity Control

"For nearly two decades airlines have been implementing revenue management systems to improve revenue results without considering the risks assumed or the consequences" (Lancaster, 2003, p. 158).

Lancaster (2003, p. 159) complains that the core proposition of capacity control is a speculative strategy but capacity control "to date has focused solely on the reward side of managing seat inventory, thus creating hidden risks." In his paper, Lancaster (2003) does not incorporate risk-aversion into a capacity control model; using a sensitivity analysis, however, he emphasizes that revenue management can help to achieve more financially stable results.

Two major supporting arguments for the assumption of risk-neutrality are quoted frequently. (1) Given both costless insurance markets to convert a stochastic income stream to the corresponding stream of expected values and perfect capital markets to convert a deterministic income stream to the most preferred maintaining the same present value, maximizing expected revenue is straightforward for a (non-risk-loving) firm (see e.g. Kennedy et al., 1994). (2) Since the booking process is repeated over hundreds or thousands of problem instances, the impact of a single realization is low, and the law of large numbers ensures that long-term average revenues are maximized by a good risk-neutral policy.

If conditions (1) and (2) do not hold in the scenario considered, risk-averse approaches are required. A company experiencing cash-flow struggles might prefer the certainty of some immediate medium revenue to the mere potential for high revenue in the future if it has only limited access to the capital market. Small companies with fewer repetitions of the booking process, little available capital, and relatively high fares might not tolerate as much financial risk as recommended by an optimal, risk-neutral capacity control policy. And even if it were optimal for the company to behave in a risk-neutral way, the product managers in charge might have other objectives e.g. if they stand to receive a bonus when a certain revenue target is met.

The theoretical need for a risk-averse model is supported in real-life revenue management. Levin et al. (2005) report on an event promoter organizing only a few large events per year. Owing to the infrequent problem instances, a single poor realization can have severe financial consequences, and the first priority might be to recover the fixed costs for rental of a stadium or concert hall. One can imagine a similar effect in clearance sales of high-value products. Furthermore, Weatherford (2004) states that he was explicitly asked by a consultant of small airlines whether the basic, risk-neutral, capacity control approaches were suitable for a risk-averse airline. Another consultant, Tom Shelton from TNS Consulting, reported to the author that his clients felt uncomfortable with the protection levels of their implemented (risk-neutral) static capacity control model. The recommended values were too aggressive. Too many seats were reserved for high-fare demand. Thus, they manipulated the probabilities of the forecasts in order to decrease the levels to magnitudes they felt comfortable with.

Bitran and Caldentey (2003, p. 226) also highlight the fact that most product managers in charge of revenue management policies present some degree of risk-aversion. By failing to suggest mechanisms for reducing unfavorable revenue levels, traditional risk-neutral capacity control models fall short of meeting the needs of a risk-averse planner.

Seen from a different perspective, the investigation of risk-averse capacity control can also be useful for a better understanding of the risks involved with capacity control. In the literature on airline risk management, revenue management is frequently categorized as an internally driven financial risk factor (Zea, 2002). Lancaster (2003, p. 159) highlights that "there is a need to develop risk awareness in airline revenue management."

Hence, it "would be interesting to measure the impact of adding risk-aversion to the revenue management formulation" (Bitran and Caldentey, 2003, p. 226). At this point, we see a gap in the current literature on capacity control. Although there have been some approaches to incorporate risk-aversion into the broader context of revenue management, there is no analytic solution for the basic capacity control model from the perspective of a risk-averse, expected utility maximizing, decision-maker.

1.2 Risk-Aversion in Revenue Management

The impact of maximizing expected utility has been examined for a variety of sequential decision problems. Among these problems are optimal stopping problems (e.g. Hall et al., 1979, and Müller, 2000), inventory models (e.g. Bouakiz and Sobel, 1992, Avila-Godoy and Fernandez-Gaucherand, 2001, Chen et al., 2004), reservoir management (Kerr et al., 1998), and underground construction planning (Likhitruangsilp and Ioannou, 2004). Despite the very rich literature on revenue management, however, to the author's knowledge very few papers deal with risk-aversion in the broader context of revenue management, and fewer still maximize the expected utility of the decision-maker. We will summarize the main papers in this area very briefly.

The risk-averse expected utility maximizing newsvendor has been extensively studied in the literature. Eeckhoudt et al. (1995) show that the optimal order quantity for such a newsvendor is lower than the order quantity of his risk-neutral counterpart and decreasing with increasing risk-aversion. Other papers that examine price and order quantity decisions for an expected utility maximizing newsvendor are Baron (1973), Lau (1980), Agrawal and Seshadri (2000), Schweitzer and Cachon (2000), and Ibarra-Salazar (2003).

Chen et al. (2004) and Caldentey and Wein (2005) maximize expected utility of revenue for pricing approaches with non-perishable products and replenishment. Chen et al. (2004) consider an inventory and pricing model maximizing total expected utility from consumption in each time period given an additive exponential utility function. Caldentey and Wein (2005) formulate a diffusion control problem for the case of an expected utility maximizing manufacturer, who sells products over two different channels. The optimal production level, the price for long-term contracts with deterministic demand, and the acceptance or rejection of stochastic offers on an electronic spot market must all be decided upon.

The papers of Feng and Xiao (1999), Levin et al. (2005), and Lim and Shanthikumar (2007) consider pricing decisions for perishable products given a risk-averse decision-maker. Feng and Xiao (1999) decide on the timing of a single switch between two predetermined prices to maximize expected revenue while penalizing the change in variance of the revenue stream over time. Levin et al. (2005) introduce a dynamic pricing model that permits the control of the loss-probability, i.e. the probability that total revenue will fall below a minimum acceptable level. Lim and Shanthikumar (2007) examine the equivalence between a robust dynamic pricing problem and a single-resource pricing model that maximizes expected exponential utility; no structural properties of an optimal policy are proven.

Mitra and Wang (2003, 2005) investigate network revenue management in a specific traffic engineering model for bandwidth provisioning and route selection. The goal is to find an appropriate volume of traffic to admit for each customer class and route (there is no time dimension). Risk-aversion is incorporated by a mean-risk analysis, which maximizes the weighted sum of the expected revenue and the standard deviation of revenue. Conditions under which the optimization problem is an instance of convex programming are obtained.

Ball and Queyranne (2006) consider a competitive analysis of a basic capacity control model. As a performance measure, they use the competitive ratio, which guarantees on a certain level of performance compared to a clairvoyant optimal policy.

The author is aware of only one other paper on risk-aversion in capacity control, a heuristic approach suggested by Weatherford (2004). He extends a well-known heuristic approach for the risk-neutral case, EMSR (expected marginal seat revenue), to a concept called EMSU (expected marginal seat utility). In this extension, the price of a ticket is replaced by the utility of its price; see Sect. 7.2.4. By means of examples, he finds that the heuristic can have a significant impact on the expected utility and revenue performance. In addition, it is reported to increase the probability of hitting certain revenue thresholds.

1.3 Chapter Organization

Our goal is (1) to introduce a new expected revenue maximizing capacity control model that accounts for cancelations and no-shows and evolves in a random environment and (2) to analyze the two most frequently addressed basic single-resource capacity control models for an expected utility maximizing decision-maker. We thereby focus on the use of an exponential atemporal utility function.

To pursue these goals step by step, the thesis consists of three parts.

The first part provides background on Markov decision processes and expected utility theory. We briefly introduce some notation before we summarize results both on finite and infinite horizon Markov decision processes (MDPs). In particular, we consider infinite horizon MDPs in a random environment and with an absorbing set. Due to the relatively short planning horizon in capacity control settings, discounting is usually not considered. That is why we focus on the total reward criterion in Chap. 2. The results are needed to examine capacity control models that aim at a maximization of expected revenue, i.e. a risk-neutral decision-maker. In order to model a risk-averse decision-maker, we start Chap. 3 with a brief review of expected utility theory for static as well as sequential (finite horizon) decision making. It is assumed that the decision-maker evaluates the outcomes of his decision with an increasing utility function and acts such that expected utility is maximized. For static decision problems with a one-dimensional outcome, we introduce the measure of absolute risk-aversion that can also be used to compare the degree of risk-aversion between decision-makers. Special utility functions, which are later used in examples, are introduced. Under certain conditions, sequential decision problems can be reduced to a static choice of a policy. In combination with a Markov decision process, the standard approach is to then use an additive time-separable utility function. Furthermore, we discuss the use of an atemporal utility function and state some arguments in favor of an exponential shape.

The second part of the thesis deals with risk-neutral, expected revenue maximizing, single-resource capacity control models. In Chap. 4, we introduce a capacity control model in a random environment that incorporates all eleven assumptions mentioned above except ix). Since it allows for cancelations and no-shows, the number of reservations one might sell is not limited by the number of seats. We also incorporate additional external factors that may have some impact on the request arrivals. In a capacity control model with a given number of periods until departure, such an external factor might simply be the time until departure. Other examples for external factors will be given. We make assumptions that guarantee that nested protection levels will provide a suitable control mechanism. The chapter is based on the ideas in Barz and Waldmann (2006). The dynamic and the static capacity control problem, the two textbook models for basic single-resource capacity control, are summarized in Chap. 5. These two models are special cases of the capacity control model in a random environment. Additional properties of the structure of an optimal control policy are discussed and the well-known EMSR heuristics are presented.

The third part introduces the concept of risk-aversion to capacity control models. The static and the dynamic capacity control model are recapitulated from the perspective of an expected utility maximizing decision-maker with both an additive time-separable utility function and an atemporal utility function. Again, our focus is on structural results of optimal controls.

The case of an additive time-separable utility function is discussed in Chap. 6. Given this preference structure, nested protection levels are suitable for the dynamic model but not for the static model. We illustrate our results with numerical examples. In addition, we discuss the impact of increasing risk-aversion on the preferred policy.

Similar to the investigation in Chap. 6, we examine the structure of optimal controls given an atemporal utility function in Chap. 7. By means of examples, we show that most structural results known from the expected revenue maximizing policy generally do not hold for a policy that maximizes expected atemporal utility. In the case of an exponential utility function, however, many structures of an expected atemporal utility maximizing policy can be proven for both the static and the dynamic capacity control problems. The results are essentially the same as those given in Barz and Waldmann (2007). Furthermore, we can show that if there is a policy that is optimal for all sufficiently small values of γ, this policy is the optimal policy of a risk-neutral decision-maker. For the static model, straightforward extensions of the EMSR heuristics are introduced to account for constant absolute risk-aversion. The results of a simulation study in a static capacity control setting with a risk-averse decision-maker are presented to underpin our findings. We compare our results with those of Weatherford's EMSU heuristic. Parts of this simulation study were published in Barz (2006).

In Chap. 8, we examine how certain structures of an expected revenue maximizing policy also carry over to the expected atemporal exponential utility maximizing case for more advanced models, such as the capacity control model under a general discrete choice model of consumer behavior. This model was introduced by Talluri and van Ryzin (2004a) in the risk-neutral case. The chapter reformulates and extends the ideas of Barz and Schön (2006).

Chapter 9 contains our summary and conclusion. In addition, directions for future research are proposed.

Part I

Basic Principles

The next two chapters cover the foundations of sequential decision making. In Chap. 2, we introduce finite and infinite horizon Markov decision processes for maximizing total (undiscounted) reward. In Chap. 3, we generalize the finite horizon case to account for an expected utility maximizing decision-maker. In particular, we focus on the maximization of expected exponential utility of the total reward. The main intention is to clarify notation and to state results that will be used in the following chapters. But first we turn to some general notation that is used throughout this work.

General Notation

We use $\mathbb{Z}$ ($\mathbb{N}_0, \mathbb{N}$) to denote the set of all (non-negative, positive) integers. A real-valued function g defined on an arbitrary set $\mathfrak{H}$ is said to be increasing if $x \leq x'$ implies $g(x) \leq g(x')$, it is said to be decreasing if $x \leq x'$ implies $g(x) \geq g(x')$ for all $x, x' \in \mathfrak{H}$.

In addition, the notation $\|\cdot\|$ will be used to denote the supremum norm, i.e.

$$\|g\| := \sup_{x \in \mathfrak{H}} |g(x)|$$

for any arbitrary set $\mathfrak{H}$ and $g : \mathbb{R} \to \mathfrak{H}$.

We use

$$\Delta g(x) := g(x) - g(x-1) , \qquad x \in \mathbb{N} ,$$

to denote the absolute increase of $g : \mathbb{N}_0 \to \mathbb{R}$. In addition, for all $g : \mathbb{N}_0 \to \mathbb{R}\backslash\{0\}$, we use

$$\Gamma g(x) := \frac{g(x)}{g(x-1)} , \qquad x \in \mathbb{N} ,$$

to denote the relative increase of g.

A function $g : \mathbb{Z} \to (0, \infty)$ is called log-convex, if $\ln g$ is convex, or, equivalently, if

$$g(x+1)^2 \leq g(x+2)g(x)$$

holds for all $x \in \mathbb{Z}$. Rearranging this inequality provides another characterization of log-convexity, namely that $\Gamma g(x)$ is increasing in x.

Log-convex functions have the nice property of being closed under both addition and multiplication (Roberts and Varberg, 1973, p. 19). The fact that log-convex functions are closed under multiplication is an immediate consequence of the characterization $g(x+1)^2 \leq g(x+2)g(x)$. To verify closure under addition, note that

$$\begin{aligned}
&(g_1(x+1) + g_2(x+1))^2 \\
&= g_1(x+1)^2 + g_2(x+1)^2 + 2g_1(x+1)g_2(x+1) \\
&\le g_1(x)g_1(x+2) + g_2(x)g_2(x+2) + 2\sqrt{g_1(x)g_1(x+2)g_2(x)g_2(x+2)} \\
&\le (g_1(x+2) + g_2(x+2))\,(g_1(x) + g_2(x)) \ ,
\end{aligned}$$

where the first inequality follows from the log-convexity of g_1 and g_2 and the second from the binomial theorem.

Finally, we sometimes write

$$x^+ := \max\{0, x\} \ , \qquad x \in \mathbb{R} \ ,$$

to simplify the notation.

2

Markov Decision Processes and the Total Reward Criterion

In this chapter, we summarize results both on finite and infinite horizon Markov decision processes (MDPs) in a random environment and with an absorbing set. Finite horizon models are used in Chap. 5. In addition, they serve as a starting point for the discussion on sequential utility maximizing decision problems in Chap. 3. The results on infinite horizon models are applied in 4. Since discounting is generally not considered in capacity control models, we focus on the expected total reward criterion.

A comprehensive introduction to Markov decision processes is provided e.g. by Puterman (1994) or White (1993a). General foundations of stochastic dynamic programming can be found in Hinderer (1970).

2.1 Finite Horizon Markov Decision Processes

A finite horizon MDP describes a stochastic system that is observed at discrete times $n = 0, \dots, N$. If at time n system state x_n from the state space $\mathfrak{X}$ is observed, a decision-maker chooses an action a_n among the admissible actions $\mathfrak{A}_n(x_n)$. This action results in an immediate one-stage reward $r_n(x_n, a_n)$ and a transition to system state x_{n+1} at time $n+1$ with probability $p_n(x_n, a_n, x_{n+1})$. At time $n = N$ a terminal reward $V_N(x_N)$ is gained and the evolution is stopped.

Thus, a finite horizon MDP consists of a tuple $(N, \mathfrak{X}, \mathfrak{A}_n, p_n, r_n, V_N, \alpha)$ with

- planning horizon $N \in \mathbb{N}_0$;
- countable state space $\mathfrak{X}$;
- countable action spaces $\mathfrak{A}_n$, where $\mathfrak{A}_n(x) \subseteq \mathfrak{A}_n$ is the non-empty finite set of all admissible actions in state $x \in \mathfrak{X}$ at time $0 \leq n < N$, $\mathfrak{A} = \bigcup_{n=0,\dots,N} \mathfrak{A}_n$; the constraint set is $\mathfrak{K}_n := \{(x, a) \mid x \in \mathfrak{X}, a \in \mathfrak{A}_n(x)\}$;
- transition laws $p_n : \mathfrak{K}_n \times \mathfrak{X} \to [0, 1]$, which represent the probability $p_n(x, a, x')$ for a transition from state $x \in \mathfrak{X}$ to state $x' \in \mathfrak{X}$ given

action $a \in \mathfrak{A}_n(x)$ at time $n \in \{0, 1, \ldots, N-1\}$ (given x, n, and a, $(p_n(x, a, x'), x' \in \mathfrak{X})$ is a counting density on $\mathfrak{X}$);
- one-stage reward functions $r_n : \mathfrak{K}_n \to \mathbb{R}$, which represent the reward $r_n(x, a)$ for choosing action a in state x at time n, $|r_n(x, a)| \leq \bar{r}$ for some $\bar{r} < \infty$;
- terminal reward $V_N : \mathfrak{X} \to \mathbb{R}$, which represents the reward $V_N(x)$ for ending in state x at time N, $|V_N(x)| \leq \bar{V}$ for some $\bar{V} < \infty$;
- one-stage discount factor $0 < \alpha \leq 1$.

A decision rule specifies the action to be taken by the decision-maker. In general, this decision rule can be history-dependent and/or randomized. But due to our assumptions, we can restrict ourselves to deterministic Markovian decision rules.

Deterministic Markovian decision rules are special decision rules $f_n : \mathfrak{X} \to \mathfrak{A}_n$, which specify the action $f_n(x) \in \mathfrak{A}_n(x)$ to be taken in state x at time n. We let $\mathfrak{F}_n$ denote the set of all deterministic Markovian decision rules at time n. An N-stage deterministic Markovian policy $\pi = (f_0, \ldots f_{N-1}) \in \Pi = \mathfrak{F}_0 \times \cdots \times \mathfrak{F}_{N-1}$ then specifies the decision rule $f_n \in \mathfrak{F}_n$ to be used at time n. Instead of $\mathfrak{F}_0 \times \cdots \times \mathfrak{F}_{N-1}$ we write $\mathfrak{F}^N$ for short, if $\mathfrak{F} = \mathfrak{F}_0 = \cdots = \mathfrak{F}_{N-1}$.

In this section, we assume that the decision-maker aims at a maximization of the total reward $R := \sum_{n=0}^{N-1} r_n(x_n, a_n) + V_N(x_N)$ (i.e. $\alpha = 1$). He can pursue this goal by choosing suitable actions $a_0, a_1, \ldots, a_{N-1}$ in states $x_0, x_1, \ldots, x_{N-1}$, i.e. by choosing a policy $\pi \in \Pi$. Note, however, that the system evolves stochastically. The system states $x_0, x_1, \ldots, x_{N-1}, x_N$ are realizations of random variables $X_0, X_1, \ldots, X_{N-1}, X_N$. Thus, the total reward is random, and we may look at

$$R_\pi := \sum_{n=0}^{N-1} r_n(X_n, f_n(X_n)) + V_N(X_N) ,$$

the total reward by applying policy π. To compare the performance of different policies, it is common to take $\mathbb{E}_\pi [R_\pi]$, the expectation of R_π with respect to the product measure P_π on $\mathfrak{X}^{N+1}$,

$$\begin{aligned} &P_\pi(X_0 = x_0, \ldots, X_N = x_N) = \\ &\qquad P(X_0 = x_0) \cdot p(x_0, f_0(x_0), x_1) \cdot \ldots \cdot p(x_{N-1}, f_{N-1}(x_{N-1}), x_N) . \end{aligned}$$

In the context of MDPs, the initial state is usually assumed to be fixed and policies are evaluated by $V_\pi^N(x) = \mathbb{E}_\pi [R_\pi \mid X_0 = x]$, the conditional expected total reward over the N time periods starting from state x and following policy $\pi \in \Pi$.

We will call a policy π^* an optimal policy if for all $\pi \in \Pi$,

$$V_{\pi^*}^N(x) \geq V_\pi^N(x), \quad x \in \mathfrak{X} ,$$

and call

$$V^{N*}(x) = \sup_{\pi \in \Pi} V_\pi^N(x) = \max_{\pi \in \Pi} \mathbb{E}_\pi \left[R_\pi \mid X_0 = x \right]$$

the value function of the MDP.

One can show the following (see e.g. Puterman, 1994, p. 92):

Theorem 2.1. $V^{N*} \equiv V_0$ *is the unique solution to the optimality equation*

$$V_n(x) = \max_{a \in \mathfrak{A}_n(x)} \left\{ r_n(x,a) + \sum_{x' \in \mathfrak{X}} p_n(x,a,x') V_{n+1}(x') \right\}, \qquad (2.1)$$

which can be obtained for $n = N-1, \ldots, 0$ *iteratively, starting with* V_N*. Every policy* π^* *formed by actions* $a = f_n^*(x)$*, each maximizing the right-hand side of (2.1), is optimal, i.e. leads to* V^{N*}*.*

The optimality equation (2.1) is the standard form for representing the dynamic programming recursion. But consider the following special case, sometimes called a system with observable disturbances (see e.g. Talluri and van Ryzin, 2004b, pp. 654–655): At every time period n, $n = 0, \ldots, N$, not only the system state $x_{1,n}$ from the state space $\mathfrak{X}_1$ is observed, but also some random disturbance $x_{2,n}$ from the countable set $\mathfrak{X}_2$. The system states $x_n = (x_{1,n}, x_{2,n})$, $n = 0, \ldots, N$ are realizations of random variables $X_n = (X_{1,n}, X_{2,n})$. Given the observations $x_{1,n}$ and $x_{2,n}$, a decision-maker chooses an action a_n among the admissible actions $\mathfrak{A}_n(x_{1,n}, x_{2,n})$ defined on the state space $\mathfrak{X} = \mathfrak{X}_1 \times \mathfrak{X}_2$. This action results in a (certain) one-stage reward $r_n(x_{1,n}, x_{2,n}, a_n)$. Given $x_{1,n}$, $x_{2,n}$ and a_n, the next system state is uniquely determined to be $x_{1,n+1} = b_{n+1}(x_{1,n}, x_{2,n}, a_n)$ for some given function $b_{n+1} : \mathfrak{X} \times \mathfrak{A}_n \to \mathfrak{X}_1$. The realization of the random disturbance $x_{2,n+1}$ is a realization of $X_{2,n+1}$ with $P(X_{2,n+1} = x_{2,n+1}) = \hat{p}_{n+1}(x_{2,n+1})$. At time $n = N$ a terminal reward $V_N(x_{1n}, x_{1n})$ is gained and the evolution stops.

Applying Theorem (2.1) yields that the maximum expected total reward $V^{N*} \equiv V_0$, is the unique solution of the optimality equation

$$V_n(x) = \max_{a \in \mathfrak{A}_n(x)} \left\{ r_n(x,a) + \sum_{x_2' \in \mathfrak{X}_2} \hat{p}_{n+1}(x_2') V_{n+1}(b_{n+1}(x,a), x_2') \right\},$$

with $x = (x_1, x_2)$, which can be obtained for $n = N-1, \ldots, 0$ iteratively, starting with V_N.

Introducing the operator

$$A_n v(x_1) = \sum_{x_2 \in \mathfrak{X}_2} \hat{p}_n(x_2) v(x_1, x_2) \qquad (2.2)$$

for an arbitrary real-valued function v on $\mathfrak{X}$, the optimality equation can be rewritten as

$$A_n V_n(x_1) = \sum_{x_2 \in \mathfrak{X}_2} \hat{p}_n(x_2) \max_{a \in \mathfrak{A}_n(x)} \left\{ r_n(x,a) + A_{n+1} V_{n+1}(b_{n+1}(x,a)) \right\},$$

with terminal reward $A_N V_N(x_1)$.

Note that in this way, the original dynamic programming recursion with state space $\mathfrak{X}_1 \times \mathfrak{X}_2$ is transformed to a recursion with state space $\mathfrak{X}_1$ reducing the computational complexity. The solution $A_0 V_0(x_1)$ can be thought of as the maximum expected total reward over N time periods starting in x_1 before x_2 is observed, whereas $V_0(x) = V_0(x_1, x_2)$ is the corresponding value after the observation x_2 was made at time 0.

2.2 Infinite Horizon Markov Decision Processes

An infinite horizon Markov decision process consists of a tuple $(\mathfrak{X}, \mathfrak{A}, p, r, \alpha)$ with

- countable state space $\mathfrak{X}$;
- countable action space $\mathfrak{A}$, where $\mathfrak{A}(x) \subseteq \mathfrak{A}$ is the non-empty finite set of all admissible actions in state $x \in \mathfrak{X}$ and the constraint set is $\mathfrak{K} := \{(x,a) \mid x \in \mathfrak{X}, a \in \mathfrak{A}(x)\}$;
- transition law $p : \mathfrak{K} \times \mathfrak{X} \to [0,1]$, which represents the probability $p(x,a,x')$ for a transition from state $x \in \mathfrak{X}$ to state $x' \in \mathfrak{X}$ given action $a \in \mathfrak{A}(x)$ (given x and a, $(p(x,a,x'), x' \in \mathfrak{X})$ is a counting density on $\mathfrak{X}$);
- one-stage reward function $r : \mathfrak{K} \to \mathbb{R}$, which represents the reward $r(x,a)$ from choosing action a in state x, $|r(x,a)| \leq \bar{r}$ for some $\bar{r} < \infty$;
- one-stage discount factor $0 < \alpha \leq 1$.

An MDP with an infinite planning horizon evolves over "infinite time". As in the finite horizon version, it describes a stochastic system that is observed at discrete times $n = 0, 1, \dots$. If at time n system state x_n is observed, the decision-maker chooses an action a_n among the admissible actions $\mathfrak{A}(x_n)$. This action results in a one-stage reward $r(x_n, a_n)$ and in a transition to system state x_{n+1} at time $n+1$ with probability $p(x_n, a_n, x_{n+1})$.

Similar to the finite horizon case, let us define a (deterministic Markovian) decision rule as $f : \mathfrak{X} \to \mathfrak{A}$. The set of all decision rules is $\mathfrak{F}$, and a policy $\pi = (f_0, f_1, \dots) \in \Pi = \mathfrak{F} \times \mathfrak{F} \times \dots$ specifies the decision rule $f_n \in \mathfrak{F}$ to be used at each point in time n. Instead of $\mathfrak{F} \times \mathfrak{F} \times \dots$ we write $\mathfrak{F}^\infty$ for short. In infinite horizon models, stationary policies $\pi = (f, f, \dots) = f^\infty \in \mathfrak{F}^\infty$ assume a particularly important role.

Denote by $(X_0, X_1, \dots)$ the state process of the Markov decision process and by R_π the total reward when applying policy π. Again, the initial state is usually assumed to be fixed and a policy $\pi \in \Pi$ is usually evaluated by

$$V_\pi(x) = \mathbb{E}_\pi \left[R_\pi \mid X_0 = x\right] = \mathbb{E}_\pi \left[\sum_{n=0}^{\infty} r_n(X_n, f_n(X_n)) \mid X_0 = x\right] ,$$

the conditional expected total reward with $\alpha = 1$, starting from state x following policy $\pi \in \Pi$ with respect to the product measure P_π on $\mathfrak{X}^\infty$,

$$P_\pi(X_0 = x_0, \ldots, X_n = x_n) = \\ P(X_0 = x_0) \cdot p(x_0, f_0(x_0), x_1) \cdot \cdots \cdot p(x_{n-1}, f(x_{n-1}), x_n) \, .$$

It seems straightforward to call a policy π^* an optimal policy if for all $\pi \in \Pi$,

$$V_{\pi^*}(x) \geq V_\pi(x), \quad x \in \mathfrak{X} \, .$$

Yet the limit V_π might not exist due to the infinite planning horizon.

Schäl (1975) states two rather general conditions for convergence of this infinite sum, conditions (GA) and (C). Since these conditions are quite technical and are not considered in the sequel, we will not discuss them here. We only point out that given these conditions, the total expected reward,

$$V^*(x) = \max_{\pi \in \Pi} \mathbb{E}_\pi \left[R_\pi \mid X_0 = x \right] \, ,$$

exists and is the unique solution of

$$V^*(x) = \max_{a \in \mathfrak{A}_n(x)} \left\{ r(x,a) + \sum_{x' \in \mathfrak{X}} p(x,a,x') V^*(x') \right\} \, , \tag{2.3}$$

for all $x \in \mathfrak{X}$.

To simplify notation, we introduce operators L and U as

$$Lv(x,a) = r(x,a) + \sum_{x' \in \mathfrak{X}} p(x,a,x') v(x') \tag{2.4}$$

and

$$Uv(x) = \max_{a \in \mathfrak{A}(x)} Lv(x,a) \, . \tag{2.5}$$

Using these operators, the optimality equation (2.3) can be stated as

$$V = UV \, .$$

2.2.1 Markov Decision Processes with an Absorbing Set

An MDP with an absorbing set is an infinite horizon MDP $(\mathfrak{X}, \mathfrak{A}, p, r, \alpha)$ with a structured state space $\mathfrak{X}$. The state space $\mathfrak{X}$ contains an absorbing set $\mathfrak{J}_0 \subset \mathfrak{X}$, i.e. $\sum_{x' \in \mathfrak{J}_0} p(x,a,x') = 1$, with $r(x,a) = 0$ for $x \in \mathfrak{J}_0, a \in \mathfrak{A}(x)$. That means that if the process runs into an absorbing set, it will stay there forever. Having entered the absorbing set, the rewards are equal to zero. For a general discussion of MDPs with absorbing set, see e.g. Hinderer and Waldmann (2003, 2005), or Waldmann (2006).

Note that $\mathfrak{J}_0$ may be empty and need not be unique. In this section, however, we only consider $\mathfrak{J}_0 \neq \emptyset$ and assume $\mathfrak{J}_0$ to be arbitrary but fixed.

The set $\mathfrak{J} := \mathfrak{X} - \mathfrak{J}_0$ of transient states is called the essential state space. It is called essential, because the behavior of the process is only of interest up to the entrance time into $\mathfrak{J}_0$ and not within $\mathfrak{J}_0$.

The fact that the MDP cannot leave any state within $\mathfrak{J}_0$ can be used to find conditions that ensure the convergence of $V_\pi(x)$ for $x \in \mathfrak{J}$. Let $\tau := \inf\{n \in \mathbb{N} | X_n \in \mathfrak{J}_0\} \leq \infty$ denote the entrance time into the absorbing set $\mathfrak{J}_0$, i.e. the first time the state process (X_n) is in $\mathfrak{J}_0$, having started in some state in $\mathfrak{J}$. Note that using τ, the total reward R_π is

$$V_\pi(x) = \mathbb{E}\left[\sum_{n=0}^{\tau-1} r(X_n, f(X_n)) | X_0 = x\right] .$$

Given a policy $\pi \in \mathfrak{F}^\infty$ and an initial state $x \in \mathfrak{J}$, the distribution of τ can be obtained by evaluating $P_\pi(\tau > n | X_0 = x)$, $n \in \mathbb{N}_0$ recursively.

In order to find an upper bound for $P_\pi(\tau > n | X_0 = x)$, define an operator H on $\mathfrak{B}$, the set of all bounded functions on $\mathfrak{J}$ (with respect to the supremum norm), by

$$Hv(x) := \max_{a \in \mathfrak{A}(x)} \sum_{x' \in \mathfrak{J}} p(x, a, x')v(x') ,$$

for $x \in \mathfrak{J}$ and $v \in \mathfrak{B}$. Let $H^{n+1}v = H(H^n v)$ and $H^0 v = 1$, $v \in \mathfrak{B}, n \in \mathbb{N}_0$. Then,

$$H^n 1(x) = \sup_{\pi \in \mathfrak{F}^\infty} P_\pi(\tau > n | X_0 = x) ,$$

with 1 denoting a vector with entries 1. Obviously, $\|H^n 1\|$ is an upper bound for the probability that the process has not entered the absorbing set $\mathfrak{J}_0$ at time n.

Hinderer and Waldmann (2005) show that the following equivalent assumptions ensure the existence of the total expected reward in an infinite horizon MDP with $\alpha = 1$:

(AS) $\|H^{n'} 1\| < 1$ for some $n' \in \mathbb{N}$.
(AS') $\|H^n 1\| \to 0$ as $n \to \infty$.

They prove the following theorem.

Theorem 2.2. *Given (AS) or (AS'),*

(i) The expected total reward

$$V^*(x) = \max_{\pi \in \Pi} \mathbb{E}_\pi\left[\sum_{n=0}^{\infty} r(X_n, f(X_n)) \mid X_0 = x\right] ,$$

is the unique bounded solution of the optimality equation $V = UV$,

$$V^*(x) = \max_{a \in \mathfrak{A}(x)} \left\{ r(x, a) + \sum_{x' \in \mathfrak{X}} p(x, a, x')V^*(x') \right\} , \qquad x \in \mathfrak{J} .$$

(ii) *A decision rule f is optimal if and only if f is a maximizer of LV (i.e. $UV(x) = LV(x, f(x))$ for all $x \in \mathfrak{J}$). Thus, there exists an optimal stationary (deterministic Markovian) policy $f^{\infty *}$.*

(iii) *Value iteration works, i.e. for all $v_0 \in \mathfrak{B}$ it holds that $v_n := Uv_{n-1}$, $n \in \mathbb{N}$, converges in norm to V (i.e. $\|V - v_n\| \to 0$ as $n \to \infty$).*

Given the assumption that the upper bound for the probability that the process has not entered the absorbing set $\mathfrak{J}_0$ at time n, $\|H^n 1\|$, converges to zero as n tends to infinity, value iteration can be used for finding the optimal policy and the associated expected total reward. This is equivalent to assuming that there is some $n' \in \mathbb{N}$ for which it is ensured that this upper bound of $P_\pi(\tau > n' | X_0 = x)$ is smaller than 1.

2.2.2 Markov Decision Processes in a Random Environment

An MDP in a random environment is another special infinite horizon MDP. It is assumed that there is some external process (an environment) that can be described by a Markov chain $\{E_n, n \in \mathbb{N}_0\}$ with state space $\mathfrak{E}$ and transition probabilities $\tilde{p}_{ee'}, e, e' \in \mathfrak{E}$. The state of the environment has some impact on the system behavior. For simplicity, let us assume it only has an impact on the transition probabilities and one-stage rewards. The evolution of the system, however, has no influence on the environment. In particular, the environment cannot be controlled by the decision-maker's actions. Yet the decision-maker can observe the state of the environment and knows its transition matrix $\tilde{P} = (\tilde{p}_{ee'})$.

In this setting, the decision-maker observes the system state s_n and environmental state e_n at time n. Based on this information, he chooses an action a_n among the admissible actions $\mathfrak{A}(s_n, e_n)$. This action results in an immediate reward $r(s_n, e_n, a_n)$ and in a transition to state s_{n+1} at time $n+1$ with probability $p(s_n, e_n, a_n, s_{n+1})$. Independently from this action and the state of the system, the environmental state changes to e_{n+1} with probability $\tilde{p}(e_n, e_{n+1})$.

Thus, an MDP in a random environment $(\mathfrak{S} \times \mathfrak{E}, \mathfrak{A}, \tilde{p}, p, r, \alpha)$ consists of a

- countable extended state space $\mathfrak{X} = \mathfrak{S} \times \mathfrak{E}$, which consists of the state space of the system $\mathfrak{S}$ and of the environment $\mathfrak{E}$;
- countable action space $\mathfrak{A}$, where $\mathfrak{A}(s, e) \subseteq \mathfrak{A}$ is the non-empty finite set of all admissible actions in system state $s \in \mathfrak{S}$ and environmental state $e \in \mathfrak{E}$; the constraint set is defined as $\mathfrak{K} := \{(s, e, a) \mid (s, e) \in \mathfrak{X}, a \in \mathfrak{A}(s, e)\}$;
- transition law of the environment $\tilde{p} : \mathfrak{E} \times \mathfrak{E} \to [0, 1]$, which represents the probability $\tilde{p}(e, e')$ for a transition from environmental state $e \in \mathfrak{E}$ to $e' \in \mathfrak{E}$ (for fixed e, $(\tilde{p}(e, e'), e' \in \mathfrak{E})$ is a counting density on $\mathfrak{E}$);
- transition law of the system $p : \mathfrak{K} \times \mathfrak{S} \to [0, 1]$, which represents the probability $p(s, e, a, s')$ for a transition from state $s \in \mathfrak{S}$ to state $s' \in \mathfrak{S}$ given environmental state $e \in \mathfrak{E}$ and action $a \in \mathfrak{A}(s, e)$ (given s, e and a, $(p(s, e, a, s'), s' \in \mathfrak{S})$ is a counting density on $\mathfrak{S}$);

- bounded one-stage reward function $r : \mathfrak{K} \to \mathbb{R}$, which represents the reward $r(s,e,a)$, $|r(s,e,a)| \leq \bar{r}$ for some $\bar{r} < \infty$, from choosing action a in system state s given the environment e;
- one-stage discount factor $0 < \alpha \leq 1$.

For a general discussion of MDPs in a random environment, see Waldmann (1981); for extensions and applications, see e.g. Waldmann (1983, 1984), Helm and Waldmann (1984), or Hinderer and Waldmann (2001).

By introducing the extended state $x = (s,e) \in \mathfrak{X}$, an MDP in a random environment can be seen as a special infinite horizon MDP with state space $\mathfrak{X}$, action space $\mathfrak{A}$, transition law $\tilde{p}(e,e')p(s,e,a,s')$ from $x = (s,e) \in \mathfrak{X}$ to $x' = (s',e') \in \mathfrak{X}$, one-stage reward $r(x) = r(s,e)$, and discount factor α. Therefore, the total expected reward ($\alpha = 1$) following policy π given state (s,e),

$$V_\pi(s,e) = \mathbb{E}_\pi \left[\sum_{n=0}^{\infty} r(S_n, E_n, A_n) \mid S_0 = s, E_0 = e \right] ,$$

is well defined under the same conditions as in the general infinite horizon case, and the optimality equation (2.3) can be restated as

$$V^*(s,e) = \max_{a \in \mathfrak{A}_n(s,e)} \left\{ r(s,e,a) + \sum_{e' \in \mathfrak{E}} \sum_{s' \in \mathfrak{S}} \tilde{p}(e,e')p(s,e,a,s')V^*(s',e') \right\} .$$

Of course, MDPs in a random environment can also have an absorbing set. Let us assume that the environment $\mathfrak{E}$ contains an absorbing set $\tilde{\mathfrak{J}}_0 \subset \mathfrak{E}$, i.e. $\sum_{e' \in \tilde{\mathfrak{J}}_0} \tilde{p}(e,e') = 1$, with $r(e,s,a) = 0$ for $e \in \tilde{\mathfrak{J}}_0, s \in \mathfrak{S}, a \in \mathfrak{A}(s,e)$. That means that if the environmental process runs into an absorbing set, it will stay there forever, and the rewards that can be gained from the system are equal to zero. We call the set $\tilde{\mathfrak{J}} := \mathfrak{E} - \tilde{\mathfrak{J}}_0$ of transient states the essential state space of the environment.

Denote by $\tilde{P}_{\tilde{\mathfrak{J}}}$ the substochastic matrix resulting from $\tilde{P}$ when the rows and columns of all environmental states belonging to $\tilde{\mathfrak{J}}_0$ are dropped. Then the nth power of this matrix, $\tilde{P}^n_{\tilde{\mathfrak{J}}} = (\tilde{p}^{(n)}_{\tilde{\mathfrak{J}}}(e,e'))$, multiplied by the vector 1, can be thought of as the probability that the environmental process has not yet entered the absorbing set $\tilde{\mathfrak{J}}_0$ at time n starting in $\tilde{\mathfrak{J}}$.

In the notation of MDPs with an absorbing set, $\mathfrak{J}_0 = \mathfrak{S} \times \tilde{\mathfrak{J}}_0$ and $\mathfrak{J} = \mathfrak{S} \times \tilde{\mathfrak{J}}$. The assumption

(ASE) $\quad \|\tilde{P}^n_{\tilde{\mathfrak{J}}} 1\| < 1$ for some $n \in \mathbb{N}$

then implies (AS), because given (ASE)

$$\begin{aligned}
\|H^n 1\| &= \| \max_{\pi \in \mathfrak{F}^\infty} \sum_{(s',e') \in \mathfrak{J}} P_\pi \left(S_n = s', E_n = e' \mid S_0 = s, E_0 = e\right) \| \\
&= \| \max_{\pi \in \mathfrak{F}^\infty} \sum_{s' \in \mathfrak{S}} \sum_{e' \in \tilde{\mathfrak{J}}} \tilde{p}^{(n)}_{\tilde{\mathfrak{J}}}(e,e') P_\pi \left(S_n = s' \mid S_0 = s, E_0 = e\right) \| \\
&< \max_{\pi \in \mathfrak{F}^\infty} \sum_{s' \in \mathfrak{S}} P_\pi \left(S_n = s' \mid S_0 \in \mathfrak{S}, E_0 \in \tilde{\mathfrak{J}}\right) = 1 \,.
\end{aligned}$$

Note that as a special case, a finite horizon MDP can also be seen as an MDP in a random environment with an absorbing set, where the environment corresponds to a time counter with state space $\mathfrak{E} = \{0, 1, \ldots, N, N+1\}$ and transition probabilities $\tilde{p}_{ee'} = 1$ for $e' = e + 1, e \leq N$, $\tilde{p}_{N+1,N+1} = 1$, $\tilde{p}_{ee'} = 0$ otherwise. The absorbing set is $\tilde{\mathfrak{J}}_0 = \{N + 1\}$. The random disturbances mentioned in the finite horizon setting can of course also be seen as a special random environment. Note, however, that the concept of a random environment E is much more general than that of random disturbances.

3

Expected Utility Theory for Sequential Decision Making

This chapter deals with decision problems under risk as defined in Knight (1921), i.e. the problem of choosing from a number of options, each of which could give rise to more than one possible outcome with different probabilities. As in Chap. 2, however, the main intention is to clarify notation and to state results that will be used in the following chapters, not to give a complete overview on expected utility theory. For a general introduction and further references, see e.g. the textbooks of Kreps (1988), French (1986), Gollier (2001), or Bamberg and Coenenberg (2006).

We will start with a brief introduction to expected utility theory in the context of static decision problems to highlight the general idea of expected utility and to provide a formal understanding of the term risk-aversion. We will thereby focus on the measure of absolute risk-aversion, because it can be easily used to compare the degree of risk-aversion for different decision-makers. This will enable us to analyze the impact of increasing risk-aversion in subsequent chapters. In addition, constant absolute risk-aversion – the case of an exponential utility function – turns out to simplify the analysis of sequential decision problems substantially. Special types of utility functions used in examples in Chaps. 6 to 8 are introduced.

The second part of this chapter deals with straightforward extensions of static decision models to problems with sequential decision making (given a finite planning horizon). Two special structures of utility functions are discussed, an additive time-separable and an atemporal utility function. Although the maximization of expected atemporal utility generally requires an enlarged state space, we will see that this is not the case given an exponential (or linear) utility function.

3.1 Static Decision Problems

In this section, we consider one-shot static decision problems under risk.

The different decision options are commonly represented by simple lotteries, where a lottery connects two types of information. The first is the set of all possible consequences, called outcomes $\mathfrak{O}$. An outcome $\vec{\mathfrak{l}}_m = (\mathfrak{l}_m^1, \mathfrak{l}_m^2, \ldots, \mathfrak{l}_m^k) \in \mathfrak{O}$ is a set of values for all variables that affect the well-being of a decision-maker. For simplicity, we assume that both the number of these variables k and the number of possible outcomes is finite, so $\mathfrak{O} = \{\vec{\mathfrak{l}}_1, \ldots, \vec{\mathfrak{l}}_{m_{\max}}\}$. Given this invariant set of potential outcomes, a lottery can be defined solely by its second ingredient, the vector of probabilities $\vec{p} = (p_1, \ldots, p_{m_{\max}}) \in \mathfrak{L}$. This vector can be interpreted as a probability of p_1 for the outcome $\vec{\mathfrak{l}}_1$, a probability of p_2 for the outcome $\vec{\mathfrak{l}}_2$, …, and a probability of $p_{m_{\max}}$ for the outcome $\vec{\mathfrak{l}}_{m_{\max}}$. The set of all simple lotteries is given by $\mathfrak{L} = \{\vec{p} = (p_1, p_2, \ldots, p_{m_{\max}}) \in \mathbb{R}^{m_{\max}} \mid \sum_{m=1}^{m_{\max}} p_m = 1, p_m \geq 0 \text{ for all } m = 1, \ldots, m_{\max}\}$.

We assume that probabilities and outcomes are all that matter to the decision-maker, the randomizing devices (e.g. whether a die is thrown or a coin is flipped) and their order are inconsequential to him. Every lottery $\vec{p} \in \mathfrak{L}$ can therefore be represented by a discrete random variable $\vec{\mathcal{O}}$ with values in $\mathfrak{O}$. The probabilities $P(\vec{\mathcal{O}} = \vec{\mathfrak{l}}_m)$ equal p_m for all $m = 1, \ldots, m_{\max}$.

The decision-maker is assumed to have a rational preference relation $\succeq$ over the set of lotteries $\mathfrak{L}$ with an expected utility representation, i.e. there exists a function $u : \mathfrak{O} \to \mathbb{R}$ on the set of outcomes such that for any two lotteries $\vec{p}^1 = (p_1^1, \ldots, p_{m_{\max}}^1)$ and $\vec{p}^2 = (p_1^2, \ldots, p_{m_{\max}}^2)$, we have

$$\vec{p}^1 \succeq \vec{p}^2 \sum_{m=1}^{m_{\max}} p_m^1 u(\vec{\mathfrak{l}}_m) = \mathbb{E}[u\left(\vec{\mathcal{O}}^1\right)] \geq \mathbb{E}[u\left(\vec{\mathcal{O}}^2\right)] = \sum_{m=1}^{m_{\max}} p_m^2 u(\vec{\mathfrak{l}}_m) . \quad (3.1)$$

This assumption was first proposed in Bernoulli (1738) and forms the basis of expected utility theory. Von Neumann and Morgenstern (1947) show that an expected utility representation follows from two rather appealing axioms on the decision-maker's rational preference relation, namely the continuity and the independence axioms. Savage (1954) provides an axiomatic basis to address with subjective probabilities and utility values within the expected utility framework. Generalizations to non-simple lotteries can be found e.g. in Fishburn (1970).

Note that if u^1 represents the preference relation $\succeq$ in the sense of (3.1), then a function $u^2 : \mathfrak{O} \to \mathbb{R}$ also represents $\succeq$ in this sense if and only if there exist real numbers $\beta_1 > 0$ and β_2 such that $u^2(\cdot) = \beta_1 u^1(\cdot) + \beta_2$.

Experiments frequently question the existence of an expected utility representation of human decision making, the most prominent example of which is probably given in Allais (1953). Therefore, the search for alternative models of decision making has been a field of research in economics for more than twenty years. For an overview of these alternative models, which are often referred to as "non-expected utility theories", see e.g. Machina (1987), Starmer (2000), or Schmidt (2004). From a normative point of view, however, the existence of an expected utility representation in static decision problems can be justified (Gollier, 2001, p. 14, or Eisenführ and Weber, 2003, pp. 212–217).

In the following subsections, we only consider lotteries with (random) one-dimensional monetary outcome, i.e. $k = 1$, the decision-maker's final wealth $W = \vec{\mathcal{O}} \in \mathfrak{O} = \{w_1, \ldots, w_{m_{\max}}\} \subseteq \mathbb{R}$. Consequently, we assume the utility function $u : \mathfrak{O} \to \mathbb{R}$ to be increasing.

3.1.1 The Concept of Risk-Aversion

Decision-makers might differ in their attitude towards risk. Within the theory of expected utility, different attitudes towards risk can be expressed by the shape of the utility function.

A Characterization of Risk-Aversion

An expected utility maximizing decision-maker with utility function u is called risk-averse if he prefers the expected outcome of a non-degenerate lottery to the lottery itself. Representing the lottery by the random variable W with expectation $\mathbb{E}[W]$, this is equivalent to

$$u\left(\mathbb{E}[W]\right) \geq \mathbb{E}\left[u(W)\right] \quad \text{for all } W \in \mathfrak{O} \,. \tag{3.2}$$

Jensen's inequality yields that a decision-maker with utility function u is risk-averse if and only if the utility function u is concave. In the case of equality in (3.2), he is called risk-neutral. This holds for linear utility functions. Given the inverse inequality, he is a risk-lover, which corresponds to a convex utility function.

Now suppose, the decision-maker may choose between playing the lottery with random outcome W or receiving a certain amount of money w_{cer}. The certainty equivalent of a lottery is the value of w_{cer} that makes the decision-maker indifferent between the two options. Thus, it fulfills

$$\mathbb{E}\left[u(W)\right] = u(w_{\text{cer}}) \,.$$

Consequently, the certainty equivalent can be defined as

$$w_{\text{cer}} = u^{-1}\left(\mathbb{E}\left[u(W)\right]\right)$$

if the utility function u is invertible. Choosing the lottery with the highest expected utility corresponds to choosing the lottery with the highest certainty equivalent.

If we assume that the decision-maker is risk-averse, he prefers the expected outcome of a lottery to playing the lottery. In other words,

$$u\left[\mathbb{E}(W)\right] \geq \mathbb{E}\left[u(W)\right] = u(w_{\text{cer}}) \,,$$

using (3.2). Consequently, $w_{\text{cer}} \leq \mathbb{E}[W]$ owing to the assumed monotonicity of u.

Comparative Risk-Aversion

Given a continuous utility function u, a common measure for risk-aversion is the Arrow-Pratt coefficient of absolute risk-aversion given wealth w,

$$\gamma(w) = -\frac{u''(w)}{u'(w)} . \tag{3.3}$$

This is basically a measure of the concavity, $-u''(w)$, normalized by the slope $u'(w)$ in order to make it invariant towards (equivalent) linear transformations, which was introduced by Pratt (1964) and Arrow (1965).

When comparing the preferences of two decision-makers with utility functions u^1 and u^2, decision-maker 1 is more risk-averse than decision-maker 2 in the sense of Pratt (1964) if for their corresponding coefficients of absolute risk-aversion $\gamma_1(w) \geq \gamma_2(w)$ holds for all $w \in \mathbb{R}$. This is equivalent to assuming that $u^1(w) = g(u^2(w))$ for all $w \in \mathbb{R}$ for some concave function g. In this case, the certainty equivalent of decision-maker 1 is not higher than the certainty equivalent of decision-maker 2 for all lotteries.

3.1.2 Special Utility Functions

As mentioned before, we only consider one-dimensional monetary outcomes. From (3.1), we know that decision-makers with linear utility functions maximize expected value. They are risk-neutral. Many other types of monotone increasing utility functions on wealth are discussed in the literature. In the following, we briefly introduce a utility function with a real-valued aspiration level as well as the logarithmic and exponential utility functions. In particular, we turn our attention to the exponential utility function and those of its properties that will be needed in the sequel.

Utility Functions with an Aspiration Level

A rather extreme utility function assigns a utility of 1 to outcomes higher than or equal to some aspiration level $\beta \in \mathbb{R}$ and 0 otherwise,

$$u(w) = \begin{cases} 1 & w \geq \beta \\ 0 & w < \beta . \end{cases}$$

Given this structure, maximizing expected utility is equivalent to maximizing the probability of an outcome of β or more. As an example of a decision-maker with such a utility function, consider a product manager who gets a bonus if the outcome is at least β. If the target is not reached, he receives nothing.

Note that this utility function is not differentiable. Therefore, the coefficient of absolute risk-aversion is not defined. Yet Müller (2000) shows by a limiting argument that the coefficient of absolute risk-aversion increases infinitely in β. A decision-maker with such a utility function is infinitely risk-loving for $w < \beta$ and infinitely risk-averse for $w \geq \beta$.

The Logarithmic Utility Function

The logarithmic utility function

$$u(w) = \log(w) , \qquad w > 0 ,$$

is strictly increasing and concave. A decision-maker with this utility function has decreasing absolute risk-aversion in w. The product $w\gamma(w)$, however, is constant.

Given a decision-maker with logarithmic utility, his certainty equivalent of the lottery is multiplied by a constant if all outcomes of a lottery are multiplied by the same amount. Thus, a multiplication of all possible outcomes by the same factor does not change his preferences between lotteries. Owing to this analytical tractability, the logarithmic utility function is frequently used, mainly in financial optimization problems.

The Exponential Utility Function

A decision-maker might be willing to accept the so-called "delta property" which stipulates that if an amount delta is added to all outcomes of a lottery, then he will want his certainty equivalent for the lottery to increase by the same amount. Accepting this delta property implies that the preferences towards lotteries are independent of the decision-maker's current wealth. One can show that the utility function of someone accepting the delta property must be either linear or exponential (Keeney, 1982, p. 90). The linear case implies risk-neutrality. In the case of risk-aversion, the exponential utility function is given by

$$u_\gamma(w) = -\exp(-\gamma w) , \qquad \gamma > 0 . \tag{3.4}$$

The shape of u_γ is illustrated in Fig. 3.1. A decision-maker with this utility

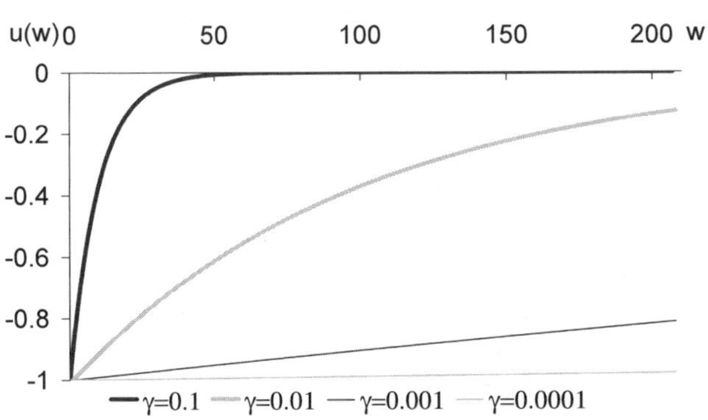

Fig. 3.1. Exponential utility function for different values of γ.

function exhibits a constant coefficient of absolute risk-aversion $\gamma(w) = \gamma$ for all values of w. The exponential utility function is strictly increasing with inverse utility function

$$u_\gamma^{-1}(w) = -\frac{1}{\gamma}\ln(-w)\ .$$

Its special form implies

$$u_\gamma(w_1 + w_2) = -\exp(-\gamma(w_1 + w_2)) = \exp(-\gamma w_2)u_\gamma(w_1)\ , \tag{3.5}$$

i.e. if the same amount w_2 is added to all outcomes of a lottery, the utilities are multiplied by $\exp(-\gamma w_2)$.

An expected utility maximizing decision-maker with an exponential utility function and small γ evaluates lotteries in terms of a trade-off between expectation and variance. Using $\exp(x) = 1 + x + x^2 + O(x^3)$ and $\ln(1 + x) = x - x^2/2 + O(x^3)$ for sufficiently small x, a Taylor expansion for the certainty equivalent of expected exponential utility yields

$$\begin{aligned} -\frac{1}{\gamma}\ln\left(\mathbb{E}[\exp(-\gamma W)]\right) &= -\frac{1}{\gamma}\ln\left(1 - \gamma\mathbb{E}[W] + \gamma^2\mathbb{E}[W^2] + O(\gamma^3)\right) \\ &= \mathbb{E}[W] - \frac{\gamma}{2}\mathrm{Var}(W) + O(\gamma^2)\ , \end{aligned} \tag{3.6}$$

with $\mathrm{Var}(W)$ denoting the variance of W. So for $\gamma \to 0$, these preferences approximate the risk-neutral objective (see Coraluppi, 1997, p. 24, or Coraluppi and Marcus, 1999). For $\gamma \to \infty$, the maximization of expected exponential utility reduces to a worst-case maximization as proven by Coraluppi (1997, p. 43) and Coraluppi and Marcus (1999) using a modified version of the Varadhan-Laplace lemma.

Exponential utility functions are the most widely used non-linear utility functions (Corner and Corner, 1985). According to Howard (1988, p. 689), they "satisfactorily treat a wide range of individual and corporate risk preferences." Furthermore, Kirkwood (2004) shows that in most cases, an appropriately chosen exponential utility function is a very good approximation for general utility functions.

Another advantage of the exponential utility function is that the preferences of such a decision-maker are completely determined by the value γ. This makes it easy to apply. For individual decision-makers simple preference comparisons between a lottery and some sure gain can be used to find γ; see e.g. Eisenführ and Weber (2003, p. 180). For business decision-makers Kirkwood (2004) states that values of 0.1 times the decision-maker's asset available within the planning horizon, the "planning asset position", seem to be realistic estimates for $1/\gamma$. A small study with only three companies in Howard (1988) indicates that a first estimate for $1/\gamma$ might be obtained by financial measures. According to this study, the values of $1/\gamma$, given by the decision-maker in charge, equal roughly 6 percent of net sales, one sixth of equity, or 125 percent of net income.

3.2 Sequential Decision Problems

Finite horizon sequential (or dynamic) decision problems deal with a sequence of choices that have to be made over time. To be more precise, a stochastic system is observed at discrete times $n = N, \ldots, 0$. (Note that we count time backwards.) If at time n system state $x_n \in \mathfrak{X}$ is observed, a decision-maker has to opt for one of the available actions (choices) $a_n \in \mathfrak{A}_n(x_n)$. This action results in an immediate one-stage reward $r_n(x_n, a_n)$ and a transition to system state x_{n-1} at time $n-1$ with probability $p_n(x_n, a_n, x_{n-1})$. At time $n = 0$ a reward $V_0(x_0)$ is gained and the evolution is stopped.

In contrast to static decision problems, dynamic problems evolve over time. The decision-maker is faced with new information before he has to make another choice. We introduced MDPs in Chap. 2 to describe exactly this kind of decision process. In Chap. 2, however, we only considered the maximization of total expected rewards. Now we turn to the more general problem of finding the policy that is most preferred by the decision-maker. This might simplify to the maximization of total expected rewards (if the decision-maker is risk-neutral and does not care about when the reward is realized), but in general, the most preferred choices might be different.

The standard manner of handling dynamic choice problems over reward streams is to represent the decision-maker's preferences by a utility function on the vector of one-stage rewards $\vec{\mathfrak{r}} = (\mathfrak{r}^N, \ldots, \mathfrak{r}^1, \mathfrak{r}^0) \in \mathfrak{O}$ with $\mathfrak{r}^n = r_n(x_n, a_n)$ for $n \geq 1$ and $\mathfrak{r}^0 = V_0(x_0)$. The sequence of choices at times $N, \ldots, 1$ induces a probability measure on the vector of rewards. A policy, i.e. a contingent plan for making choices, is then evaluated by the induced expected utility. This means that the sequential decision problem is reduced to a static choice of a policy. This view on sequential decision problems is also used in roll-back procedures as introduced in Eisenführ and Weber (2003, pp. 240–242) or von Nitzsch (2002, pp. 214–216). General decision problems of this kind are analyzed in Kreps (1977a), (1977b), (1978), Schäl (1981), and Fainberg (1982).

There are two major objections to the above-mentioned approach (see e.g. Zacharias, 1993). First, the timing of the resolution of uncertainty is ignored. This aspect might be important from a normative point of view if other decisions must be made prior to the latest resolution of uncertainty possible. Models that take this temporal aspect into account were first proposed by Kreps and Porteus (1978) and Selden (1978). Kreps and Porteus (1979) discuss their model within the context of dynamic programming. A special parametrization was used in Epstein and Zin (1989) and Weil (1990). See Gollier (2001, Chap. 20) or Danthine and Donaldson (2002, Sect. 2.7.1) for an introduction.

A second objection arises if the decision-maker has to decide on how to consume the reward stream before uncertainty is resolved and the utility function is given in terms of the outcomes of these consumption decisions. Even if the preferences over consumption have an expected utility representation, there will generally be no expected utility representation capable of summarizing the induced preferences on the reward stream. This observation dates back to Markowitz (1959, Chap. 11), Mossin (1969), and Spence and Zeckhauser (1972). An introductory example is given in Kreps (1988, pp. 171–175). Smith (1998) presents a decision analysis approach that can be applied to reward streams directly when the decision-maker may borrow and lend to spread consumption optimally over time.

For simplicity, we assume that within the planning horizon considered, there are no decisions on consumption that depend on the outcome of the reward decision problem. In other words, the decision-maker does not stand to gain extra value from either early or late resolution of uncertainty. In particular, we assume that the decision-maker's preferences on different reward streams $\vec{\imath} \in \mathfrak{O} = \mathbb{R}^{N+1}$ have an expected utility representation $u : \mathfrak{O} \to \mathbb{R}$.

Still, general utility functions over time streams $u : \mathfrak{O} \to \mathbb{R}$, as introduced above, raise several difficulties in optimization problems. The marginal utility of reward in period n might be any function of past and future rewards. Thus, the use of computational procedures such as backward induction is difficult.

To make the analysis more tractable, we only consider two special structures of u, additive time-separable utility functions and atemporal utility functions. The first structure corresponds to the standard assumption for dealing with dynamic utility maximization problems over time. The second approach ignores the time aspect when evaluating reward streams. Both structures lead to simple representations composed of one-dimensional increasing utility functions $u_n : \mathbb{R} \to \mathbb{R}$.

To avoid technicalities, we impose the same restrictive assumptions on the state space, action space, one-stage reward functions, transition probabilities, and terminal reward as in the discussion on finite horizon MDPs in Sect. 2.1.

3.2.1 Additive Time-Separable Utility Functions

In his definition of Markov decision processes, Rust (1996) explicitly assumes that the decision-maker has a utility function that is additively separable, since other structures are "computationally intractable for solving all but the smallest problems" (Rust, 1996, p. 630).

A utility function u is additive time-separable if there exist $N+1$ functions $u_n : \mathfrak{O} \to \mathbb{R}, n = 0, \ldots, N$, such that

$$u(\vec{\imath}) = \sum_{n=0}^{N} u_n(\imath^n) . \tag{3.7}$$

(This property is sometimes also referred to as direct additivity; see e.g. Blackorby et al., 1998.) Given an additive time-separable utility function, preferences over lotteries only depend on the marginal probability distributions of the one-period rewards. In particular, the assumption of additive time-separability is equivalent to the following preference structure: Given a lottery with non-zero probabilities for the two outcomes

$$\vec{\wr}_{1,1} = (\wr^0_{1,1}, \ldots, \wr^n, \ldots, \wr^N_{1,1}); , \qquad \vec{\wr}_{1,2} = (\wr^0_{1,2}, \ldots, \wr^N_{1,2}) ,$$

and a second lottery with non-zero probabilities for the two outcomes

$$\vec{\wr}_{2,1} = (\wr^0_{2,1}, \ldots, \wr^n, \ldots, \wr^N_{2,1}); , \qquad \vec{\wr}_{2,2} = (\wr^0_{2,2}, \ldots, \wr^N_{2,2}) ,$$

the preferences of the decision-maker between these lotteries are independent of $\wr^n$. In other words, whatever preference the decision-maker has between the two lotteries, it must remain the same if the level $\wr^n$ in outcomes $\vec{\wr}_{1,1}$ and $\vec{\wr}_{2,1}$ is changed. For a proof, see Pollak (1967) or Keeney and Raiffa (1976, Sect. 6.5).

Because this structure is particularly simple to deal with analytically, an additive time-separable utility function is frequently assumed; see e.g. Lippman and McCall (1981, p. 242) or Varian (1992, p. 359). In this case, the maximization of expected utility in a finite horizon sequential decision problem corresponds to the solution of a finite horizon MDP as introduced in Sect. 2.1. The system states $x \in \mathfrak{X}$ determine the available choices $a \in \mathfrak{A}(x)$; the next system state x' is given by transition probabilities $p_n(x, x')$. The only difference to expected total reward maximization (with rewards $r_n(x, a)$) is that one-stage rewards have to be transformed by the utility function to read $u_n(r_n(x, a))$. Accordingly, the terminal reward is $u_0(V_0)$.

Given the notation introduced in Sect. 2.1, we can use the results of Theorem 2.1 to determine the optimal policy and the maximum expected utility

$$V^{\text{add}*}(x) = \max_{\pi \in \mathfrak{F}^N} \mathbb{E}_\pi \left[\sum_{n=1}^{N} u_n\left(r_n(X_n, f_n(X_n))\right) + u_0\left(V_0(X_0)\right) \mid X_N = x \right] .$$

In particular, $V^{\text{add}*} \equiv V^{\text{add}}_N$ is the unique solution to the optimality equation

$$V^{\text{add}}_n(x) = \max_{a \in \mathfrak{A}_n(x)} \left\{ u_n(r_n(x, a)) + \sum_{x' \in \mathfrak{X}} p_n(x, a, x') V^{\text{add}}_{n-1}(x') \right\} , \tag{3.8}$$

which can be obtained for $n = 1, \ldots, N$ iteratively, starting with the terminal reward $u_0(V_0)$. Every policy $\pi^{\text{add}*}$ formed by actions $a = f^{\text{add}*}_n(x)$, each maximizing the right-hand side of (3.8), is add-optimal, i.e. leads to $V^{\text{add}*}$.

Keep in mind that time is counted backwards and that u_n is defined on one-stage rewards $\wr^n$ for $n = 0, \ldots, N$. The latter is why we assume $u_n(\wr)$ to be increasing in $\wr$ for all $n = 0, \ldots, N$.

Without loss of generality, $u_n(0) = 0$ for all n. This assumption is not restrictive, since linear transformations of the utility function have no impact on the represented preferences. Stated differently, if all one-stage rewards are transformed from $u_n(r_n(x, a))$ to $\beta_1 u_n(r_n(x, a)) + \beta_{2,n}$, for $\beta_1 > 0$ and $\beta_{2,n} \in \mathbb{R}$ for all n, i.e. the utility function u is linearly transformed to $\tilde{u} = \beta_1 u + \sum_{n=0}^{N} \beta_{2,n}$, the optimal policy does not change.

Varying the shape of the one-period utility functions u_n has an effect on the resistance to intertemporal substitution. For example, take concave $u_n = u_0$ for all $n = 1, \ldots, N$. Then the decision-maker has a desire for a smooth reward stream over time and he would prefer an equal reward within each period to any other reward stream with the same total value. Thus, a reward stream of e.g. $\vec{\wr}_1 = (50, 50)$ would be preferred to $\vec{\wr}_2 = (0, 100)$. Given a convex utility function u, a smooth reward stream would be the worst-case scenario. This property is discussed extensively in the literature on consumption and saving; see e.g. Gollier (2001, Sect. 15.3).

As mentioned before, concave one-period utility functions also result in risk-averse behavior within each period. In other words, the concavity of u induces intertemporal preferences and risk-aversion at the same time. Any degree of risk-aversion necessarily involves an intertemporal substitution effect as well.

Clearly, a risk-neutral decision-maker with $u_n(\wr) = \wr$ for all $\wr \in \mathbb{R}$ and $n = 0, \ldots, N$ maximizes total reward as discussed in Sect. 2.1.

3.2.2 General Atemporal Utility Functions

If either the time between two decisions is negligible and the planning horizon is short or the monetary value of the whole reward stream is paid at the end of the planning horizon, it is straightforward to assume that the decision-maker is indifferent between reward streams of the same sum

$$w = \sum_{n=0}^{N} \wr^n .$$

This corresponds to an atemporal structure of the decision-maker's utility function,

$$u(\vec{\wr}) = u_0\Big(\sum_{n=0}^{N} \wr^n\Big) = u_0(w) , \tag{3.9}$$

if the initial wealth is 0. As before, we assume $u_0(w)$ to be increasing in w.

For general utility function u_0, this objective is non-separable, since the utility of one-stage rewards depends on the rewards realized in all periods. Müller (2000) and Hall et al. (1979) use such a utility function in an optimal stopping problem. For a state-of-the-art review of MDPs in the context of expected atemporal utility maximization, see Liu (2005).

Generally, sequential decision problems that aim at a maximization of expected atemporal utility can be solved by a Markov decision process if the state space is enlarged by another variable w, the accumulated rewards up to the current period; see White (1987). To be more precise, let $\mathfrak{X}$ denote the original state space and $\mathfrak{W}$ the countable set of all possible accumulated wealth levels within the planning horizon, $\sum_{n'=0}^{n} \wr^{n'}$, $0 \leq n \leq N$. The maximization of expected atemporal utility in a finite horizon sequential decision problem will then correspond to the solution of a finite horizon MDP as introduced in Sect. 2.1 with state space $\mathfrak{X} \times \mathfrak{W}$.

The system state x and the wealth state w form the state $(x, w) \in \mathfrak{X} \times \mathfrak{W}$. The available actions are $a \in \mathfrak{A}(x)$. The next state $(x', w+r_n(x,a))$ is given by the transition probabilities $p_n(x, x')$; other system states cannot be reached. There are no one-stage rewards. V_0 is added to the wealth at the end of the planning horizon. This value is then transformed by the decision-maker's utility function to yield the terminal reward $V_0^{\text{atmp}}(x, w) = u_0(w + V_0(x))$. Again, time is counted backwards.

In line with the notation of Chap. 2, (x_N, w_N), (x_{N-1}, w_{N-1}), ..., (x_0, w_0) denote the realizations of the random variables (X_N, W_N), (X_{N-1}, W_{N-1}), ..., (X_0, W_0), and we write

$$R_\pi := W_N + \sum_{n=1}^{N} r_n(X_n, f_n(X_n, W_n)) + V_0(X_0) = W_0 + V_0(X_0)$$

for the total reward by applying policy π.

Due to the finite planning horizon, we can use the results of Theorem 2.1 to determine the optimal policy and the maximum expected utility given initial state (c, w),

$$V^{\text{atmp}*}(x, w) = \max_{\pi \in \Pi} \mathbb{E}_\pi \left[u_0\left(R_\pi\right) \mid (X_N, W_N) = (x, w) \right] .$$

In particular, $V^{\text{atmp}*} \equiv V_N^{\text{atmp}}$ is the unique solution to the optimality equation

$$V_n^{\text{atmp}}(x, w) = \max_{a \in \mathfrak{A}_n(x)} \left\{ \sum_{x' \in \mathfrak{X}} p_n(x, a, x') V_{n-1}^{\text{atmp}}(x', w + r_n(x, a)) \right\} , \qquad (3.10)$$

which can be obtained for $n = 1, \dots, N$ iteratively, starting with the terminal reward $V_0^{\text{atmp}}(x, w) = u_0(w + V_0(x))$. Every policy $\pi^{\text{atmp}*}$ formed by actions $a = f_n^{\text{atmp}*}(x, w)$, each maximizing the right-hand side of (3.10), is atmp-optimal, i.e. leads to $V^{\text{atmp}*}$.

This approach is rarely used for general utility functions, presumably because of the enlarged state space. Exceptions are e.g. Hall et al. (1979) and Kerr (1999). Yet for the special case of a utility function with an aspiration level, this approach is well-known as the target-level criterion in Markov decision processes; see e.g. White (1993b), Bouakiz and Kebir (1995), Yu (1998), Wu and Lin (1999), Ohtsubo and Toyonaga (2002), and Waldmann (2006).

Note that if the one-stage rewards are not uniquely determined by the current state and action, the maximization of $\mathbb{E}_\pi \left[u_0(W_0 + V_0(X_0))\right]$ for given X_N and W_N can be achieved by the calculation of

$$V_n^{\text{atmp}}(x, w) = \max_{a \in \mathfrak{A}_n(x)} \left\{ \sum_{x' \in \mathfrak{X}} p_n(x, w, a, x', w') V_{n-1}^{\text{atmp}}(x', w') \right\} , \qquad (3.11)$$

with terminal rewards $V_0^{\text{atmp}}(x,w) = u_0(w + V_0(x))$. We did not pursue this general approach in (3.10) in order to highlight the structural difference to the additive time-separable approach and to stick to the sequential decision problem stated above. However, we will need this flexibility in Chap. 8. Equation (3.10) forms a special case with $p_n(x,w,a,x',w') = p_n(x,a,x')$ for $w' = w + r_n(x,a)$ and $p_n(x,w,a,x',w') = 0$ else.

Again, a risk-neutral decision-maker with $u_0(w) = w$ for all $w \in \mathbb{R}$ results in the total reward criterion discussed in Sect. 2.1. In this case, the accumulated wealth need not be part of the state space. This can be seen easily by induction on n. Starting with $V_0^{\text{atmp}}(x,w) = V_0(x) + w$, one can go through (3.10) iteratively for $n = 1,\ldots,N$ and state the optimality equation as

$$\begin{aligned} V_n^{\text{atmp}}(x,w) &= \max_{a \in \mathfrak{A}_n(x)} \left\{ \sum_{x' \in \mathfrak{X}} p_n(x,a,x') V_{n-1}^{\text{atmp}}(x', w + r_n(x,a)) \right\} \\ &= \max_{a \in \mathfrak{A}_n(x)} \left\{ r_n(x,a) + \sum_{x' \in \mathfrak{X}} p_n(x,a,x') V_{n-1}^{\text{atmp}}(x', w) \right\} \\ &= \max_{a \in \mathfrak{A}_n(x)} \left\{ w + r_n(x,a) + \sum_{x' \in \mathfrak{X}} p_n(x,a,x') V_{n-1}(x') \right\} . \end{aligned}$$

Clearly, the added variable w is superfluous for the maximization. The optimal actions are the same as the ones obtained by (2.1).

3.2.3 Atemporal Exponential Utility Functions

We have just seen that for a risk-neutral decision-maker with atemporal expected utility representation and $u_0(w) = w$ for all $w \in \mathbb{R}$, the general optimality equation (3.10) can be reduced to a form in which the accumulated wealth can be discarded in the state space. This holds because $u_0(w_1 + w_2) = u_0(w_1) + u_0(w_2)$ is trivially true for all linear utility functions with $u_0(0) = 0$, independent of the values w_1 and w_2.

As mentioned before, the only other utility function that evaluates different alternatives independent of the decision-maker's current wealth is the exponential utility function $u_\gamma(w) = -\exp(-\gamma w)$ for all $w \in \mathbb{R}$ with constant absolute risk-aversion of γ. Because of its multiplicative structure, given in (3.4), we can show the following. Starting with $V_0^{\text{atmp}}(x,w) = \exp(-\gamma w) u_\gamma(V_0(x))$ and

$$\begin{aligned} V_1^{\text{atmp}}(x,w) &= \max_{a \in \mathfrak{A}_n(x)} \left\{ \sum_{x' \in \mathfrak{X}} p_1(x,a,x') V_0^{\text{atmp}}(x', w + r_1(x,a)) \right\} \\ &= \max_{a \in \mathfrak{A}_n(x)} \left\{ \sum_{x' \in \mathfrak{X}} p_1(x,a,x') \exp(-\gamma r_1(x,a)) V_0^{\text{atmp}}(x', w) \right\} , \end{aligned}$$

one can go through (3.10) iteratively for $n = 2, \dots, N$ and state the optimality equation as

$$V_n^{\text{atmp}}(x, w) = \max_{a \in \mathfrak{A}_n(x)} \left\{ \sum_{x' \in \mathfrak{X}} p_n(x, a, x') \exp(-\gamma r_n(x, a)) V_{n-1}^{\text{atmp}}(x', w) \right\} .$$

Again, the added variable w is superfluous for finding the maximizing actions. If we write V^γ instead of V^{atmp} in order to underline the use of the exponential utility function, the optimality equation can be stated as

$$V_n^\gamma(x) = \max_{a \in \mathfrak{A}_n(x)} \left\{ \exp(-\gamma r_n(x, a)) \sum_{x' \in \mathfrak{X}} p_n(x, a, x') V_{n-1}^\gamma(x') \right\} . \tag{3.12}$$

Introducing

$$V^{\gamma *}(x) = \max_{\pi \in \Pi} \mathbb{E}_\pi \left[u\left(R_\pi\right) \mid X_N = x \right] \in [-1, 0) ,$$

$V^{\gamma *} \equiv V_N^\gamma$ is the unique solution to the optimality equation (3.12). The values can be obtained for $n = 1, \dots, N$ iteratively, starting with V_0^γ. Every policy $\pi^{\gamma *}$ formed by actions $a = f_n^{\gamma *}(x)$, each maximizing the right-hand side of (3.12), is γ-optimal, i.e. leads to $V^{\gamma *}$.

For a direct derivation of (3.12), see Howard and Matheson (1972). MDPs with an expected atemporal exponential utility maximizing criterion are frequently referred to as risk-sensitive MDPs. In this work, we only consider (observable) finite horizon problems with this exponential (undiscounted) total cost criterion. This was introduced by Howard and Matheson (1972) and discussed e.g. in Rothblum (1984), Coraluppi (1997), and Avila-Godoy et al. (1997). Yet most of the literature on risk-sensitive control deals with the infinite horizon case; see the review article of Marcus et al. (1997) or the Liu's dissertation (2005) and the references given there.

In parallel to the work on risk-sensitive Markov decision processes, a body of research on risk-sensitive control has emerged in the control community. An overview is given by Whittle (1990). Risk-sensitive control problems have attracted more recent attention because of the connections to robust control and differential games; see e.g. Jacobson (1973). They are also discussed in the context of artificial intelligence; see e.g. Mihatsch and Neuneier (2002) and Liu (2005).

Part II

Expected Revenue Maximizing Capacity Control

The following two chapters deal with risk-neutral single-resource capacity control models under the independent demand assumption.

In Chap. 4, the results on infinite horizon MDPs in a random environment and with an absorbing set are applied to derive structural properties of an expected revenue maximizing policy for a capacity control model. This model incorporates additional external factors that may have some impact on the request arrivals using the concept of a random environment. In contrast to the major part of the capacity control literature, we do not assume a deterministic number of decision periods, but allow for an impact of the evolution of these external factors on the planning horizon. In particular, we assume that the state space of the random environment contains an absorbing set; the entrance time into the absorbing set determines the planning horizon. In addition, we relax assumption ix) and allow for both cancelations and no-shows as proposed by Subramanian et al. (1999) and Talluri and van Ryzin (2004b, pp. 155–161). Consequently, we do not limit the number of reservations one might sell by the number of available seats; we allow for overbooking. In our analysis, we focus on assumptions that guarantee an optimal policy of (generalized) protection level type. After giving some examples of external factors to demonstrate the great versatility, we close this chapter with two numerical examples to illustrate our results.

The two textbook models for basic single-resource capacity control – the dynamic and the static capacity control problem – are summarized in Chap. 5. These two models can be seen as special cases of the capacity control model in a random environment. We revise both models by stating the decision models, giving a short survey on the corresponding literature, discussing structural results, and presenting a short numerical example. In addition, the popular EMSR heuristics for the static model are introduced.

4

Capacity Control in a Random Environment

We consider a non-stop flight of an airplane with a capacity of C that is to depart after a certain time T. There are $i_{\max}$ ($i_{\max} \in \mathbb{N}$) booking classes, $i = 1, \ldots, i_{\max}$, with associated fares ϱ_i ordered such that $0 < \varrho_{i_{\max}} < \varrho_{i_{\max}-1} < \ldots < \varrho_1$. The number of booking periods in $[0; T]$ is given by some external process and might be random. In every booking period n, a customer requests a certain number of reservations $d_n \in \mathfrak{D} := \{0, 1, \ldots, d_{\max}\}$, $d_{\max} \in \mathbb{N}_0$, for seats of booking class $i_n \in \mathfrak{I} := \{0, 1, \ldots, i_{\max}\}$. ($i_n = 0$ with $\varrho_0 = 0$ denotes an artificial booking class corresponding to no customer request.) Thus, the $(n+1)$st customer request provides information on the number $d_n \in \mathfrak{D}$ of reservations (the customer is interested in) and the booking class $i_n \in \mathfrak{I}$ with reward ϱ_{i_n} that is offered for each of the d_n reservations. It must be decided how many of these requested reservations should be actually sold.

We allow for cancelations, no-shows, and overbooking. We also allow for group arrivals, but since it is known that structural properties fail for total accept/deny decisions (see e.g. Lee and Hersh, 1993, or Brumelle and Walczak, 2003), we assume that these groups can be partially accepted. Furthermore, we incorporate additional external factors – a concept similar to the state of the market in the dynamic pricing context of Aviv (2005) and stimulated by the control of queuing systems in a random environment as studied in Helm and Waldmann (1984). The entrance time to an absorbing set of the random environment determines the number of decision periods before departure. Consequently, this number is random, as in the semi-Markov decision processes studied by Brumelle and Walczak (2003) and Walczak (2001).

After a rigorous presentation of the decision model in the first section, we go on to analyze structural results of an optimal policy. The next section covers possible applications demonstrating the efficacy and scope of our results. Finally, we illustrate our results in two numerical examples.

4.1 The Decision Model

We describe the following steps in more detail:

(1) The next state of the external process is determined. This external state establishes the booking class and the number of requested reservations as well as flight departure;
(2) The controller decides on the number of requests to accept, i.e. tickets to sell;
(3) Some customers cancel their reservations; expected overbooking costs are incurred.

We end this section with a formulation of the underlying Markov decision process in order to analyze structural results of an optimal policy.

4.1.1 The Environmental Process and the Arrival of Requests

Customer requests (for reservations) are assumed to depend on the realization of an external process (Z_n) with countable state space $\mathfrak{Z}$. In particular, let stage n of the external process (Z_n) correspond to the $(n+1)$st request for d_n reservations in booking class i_n. The values d_n and i_n are realizations of random variables D_n and I_n, respectively, with joint distribution

$$\eta_{z_n}(d_n, i_n) := P(D_n = d_n, I_n = i_n \mid Z_n = z_n) ,$$

which may depend on the current external state z_n. On the other hand, the distribution of the future external state Z_{n+1} may depend on the realized number d_n and the realized booking class i_n (in addition to z_n). The distribution is given by

$$\kappa_{z_n,d_n,i_n}(z_{n+1}) := P(Z_{n+1} = z_{n+1} \mid Z_n = z_n, D_n = d_n, I_n = i_n) .$$

Note that the random variables D_n, I_n, and Z_n do not depend on the control of the process (which is in line with the independent demand assumption). Therefore, we may summarize all information about D_n, I_n, and Z_n in an environmental process (E_n) with countable state space $\mathfrak{E} = \mathfrak{D} \times \mathfrak{I} \times \mathfrak{Z}$. (E_n) can easily be verified to be a Markov chain with transition matrix $\tilde{P} = (\tilde{p}(e, e'))$, where, for each environmental state $e \in \mathfrak{E}$ and $e' = (d', i', z') \in \mathfrak{E}$, we have $\tilde{p}(e, e') = \kappa_e(z')\eta_{z'}(d', i')$.

To realize a finite number of requests (almost surely), we suppose that there is a non-empty absorbing set $\tilde{\mathfrak{J}}_0 \subseteq \mathfrak{E}$ and that the environmental process finally runs into an absorbing state $e \in \tilde{\mathfrak{J}}_0$. This event indicates the end of the booking process. Having entered such an absorbing state, neither more requests, nor cancelations occur; no additional costs or rewards may be gained. This assumption is in line with our definition of an absorbing set in Sect. 2.2.2. In the airline setting, absorbing states indicate that the next observable event is flight departure. As before, we denote the complement of $\tilde{\mathfrak{J}}_0$ in $\mathfrak{E}$ by $\tilde{\mathfrak{J}}$, the essential state space of the environmental process.

We postulate that there is some time period n^0 for which the probability of having entered $\tilde{\mathfrak{J}}_0$ is positive no matter where the process started.

(A1) There is some $n^0 \in \mathbb{N}$ such that $\|\tilde{P}_{\tilde{\mathfrak{J}}}^{n^0} 1\| < 1$.

Remember that $\tilde{P}_{\tilde{\mathfrak{J}}}$ denotes the substochastic matrix resulting from $\tilde{P}$ by dropping the rows and columns of all external states belonging to $\tilde{\mathfrak{J}}_0$ as introduced in Sect. 2.2.2. $\tilde{P}_{\tilde{\mathfrak{J}}}^{n^0} 1$ can be thought of as the probability that the external process has not yet entered the absorbing set $\tilde{\mathfrak{J}}_0$ at time n^0 starting in $\tilde{\mathfrak{J}}$.

Assumption (A1) does not seem to be a problem in capacity control, since the capacity is assumed to perish after some fixed time T. (A1) holds e.g. if there is only a finite number of requests (including the artificial requests with $i = 0$) within the planning horizon $[0, T]$.

4.1.2 Accepting Requests

Next, we turn to the booking process in more detail. Fix $n \in \mathbb{N}_0$. Let $c_n \in \mathfrak{C} := \{x \in \mathbb{Z} \mid x \leq C\}$ denote the number of non-reserved, or remaining, seats at stage n. Then, given the $(n+1)$st request of size d_n and class i_n, a decision has to be made about the number $a_n \in \mathfrak{A}(d_n) := \{0, \dots, d_n\}$ of requests to be accepted, leading to a reward $\varrho_{i_n} a_n$ and a decrease in the number of remaining seats from c_n to $c_n - a_n$.

To make up for the loss of reservations due to cancelations and no-shows, $c_n - a_n$ is not bounded below by 0. In this way, we allow for overbooking, which, however, is qualified by both a one-stage penalty cost $\psi^{\mathrm{P}}(c_n, e_n) \geq 0$, $(c_n, e_n) \in \mathfrak{J} = \mathfrak{C} \times \tilde{\mathfrak{J}}$, and a terminal cost $\psi^{\mathrm{T}}(c^0, e^0) \geq 0$ depending on the extended state $(c^0, e^0) \in \mathfrak{J}_0 = \mathfrak{C} \times \tilde{\mathfrak{J}}_0$ upon departure.

4.1.3 Cancelations, No-Shows, and Overbooking

As mentioned earlier, we also consider no-shows and cancelations of reservations. Cancelations result in an increase of the remaining capacity. In particular, the remaining capacity is assumed to increase from $c_n - a_n$ to $c_{n+1} \geq c_n - a_n$ between stages n and $n+1$ with a probability of $q^{e_n}_{c_n - a_n, c_{n+1}}$, say, which may depend on the external state e_n. Note that for the arrival process, we allowed i_n to be equal to 0. This represents a pure canceling event without any requests. We make the following technical assumption about the cancelation probabilities:

(A2) For all $e \in \tilde{\mathfrak{J}}$ and all increasing and concave functions $g : \mathfrak{C} \to \mathbb{R}$, it holds that

$$c \to \sum_{c'=c}^{C} q^e_{c,c'} g(c')$$

is increasing and concave.

Assumption (A2) is based on the well-known increasing concave order of random variables. It holds if the transition matrices $(q^e_{c,c'})$, $e \in \tilde{\mathfrak{J}}$, are stochastically increasing and concave in the sense of Shaked and Shanthikumar (1988). This needs to be checked in concrete applications. It is known to hold e.g. if the increase in capacity can be modeled by a binomial distribution; see Appendix C.

We assume that a customer who has not canceled in advance will not show up at departure with a probability of $q^{e^0}_{\mathrm{NS}}$, independent of the booking class but dependent on the environmental state $e^0 \in \tilde{\mathfrak{J}}_0$ at the time of departure. These no-shows can be considered within the terminal costs ψ^{T} by setting

$$\psi^{\mathrm{T}}(c^0, e^0) = \mathbb{E}_{c^0, q^{e^0}_{\mathrm{NS}}}\left[\psi^{\mathrm{DB}}_{e^0}(-c^0 - B)\right] ,$$

where B represents the number of no-shows. $\psi^{\mathrm{DB}}_{e^0}(x) \geq 0$ are the costs that stem from denying boarding to x passengers with reservations at environmental state $e^0 \in \tilde{\mathfrak{J}}_0$. These costs are typically 0 for $x \leq 0$ and increasing and convex in x. Since $C - c^0$ denotes the number of reservations at hand at the time of departure, the random variable B is assumed to follow a binomial distribution with parameters $C - c^0$ and $q^{e^0}_{\mathrm{NS}}$. This approach is the standard way of incorporating no-shows and was also applied e.g. in Subramanian et al. (1999) and Brumelle and Walczak (2003).

To incorporate these terminal costs into our model, we consider at each stage not only the above-mentioned rewards for accepting requests reduced by the penalty costs for ignoring the capacity restrictions of the airplane, but also the expected costs of overbooking weighted by the probability that there will be no more request events until departure. For $c \in \mathfrak{C}$, $e \in \tilde{\mathfrak{J}}$ define these aggregated costs as

$$\psi_e(c) := \psi^{\mathrm{P}}(c, e) + \sum_{c'=c}^{C} q^e_{c,c'} \sum_{e' \in \tilde{\mathfrak{J}}_0} \tilde{p}(e, e') \psi^{\mathrm{T}}(c', c') .$$

We make the following reasonable assumption on these penalty and overbooking costs.

(A3) For all $e = (d, i, z) \in \tilde{\mathfrak{J}}$
  (i) $\psi_e(\cdot) \geq 0$ is decreasing and convex,
  (ii) $\lim_{c \to -\infty} -\Delta\psi_e(c) > \varrho_i$.

Part (i) of (A3) is satisfied by the common assumption on the penalty costs and terminal costs for overbooking to be increasing and convex in the number of reservations at hand, i.e. in $C - c$, for every $e \in \tilde{\mathfrak{J}}$. In the case of no-shows as defined above, this is true if $\psi^{\mathrm{DB}}_{e^0}(\cdot)$ is decreasing and convex for each environmental state $e^0 \in \tilde{\mathfrak{J}}$ at departure (see Subramanian et al., 1999). Part (ii) is satisfied e.g. if an overbooking pad is introduced and the penalty costs of overbooking by more than this threshold are at least as high as the reward earned by accepting the highest-class customer (another very common assumption in the literature; see e.g. Subramanian et al., 1999).

Our modeling approach implies the usual assumptions that cancelation and no-show probabilities are the same for all customers and are independent of the time the reservations on hand were accepted. If these probabilities are mutually independent across customers, the equivalent charging scheme (as used in Subramanian et al., 1999) can be applied for modeling class dependent cancelation and no-show refunds. For simplicity, we do not consider any refunds in the following.

4.1.4 The Underlying Markov Decision Process

Our decision problem can be treated as a Markov decision process in a random environment with an absorbing set as indicated in Sect. 2.2.2. The infinite horizon MDP has countable state space $\mathfrak{X} = \mathfrak{C} \times \mathfrak{E}$, countable action space $\mathfrak{A} = \{0, \ldots, d_{\max}\}$, finite subsets $\mathfrak{A}(d) = \{0, \ldots, d\}$ of admissible actions in state $(c, e) \in \mathfrak{X}$, constraint set $\mathfrak{K} = \{((c,e),a) \mid (c,e) \in \mathfrak{X}, a \in \mathfrak{A}(d)\}$, transition law p from $\mathfrak{K}$ into $\mathfrak{X}$, where $p((c,e),a,(c',e')) := q^e_{c-a,c'}\tilde{p}(e,e')$, and reward function $r : \mathfrak{K} \to \mathbb{R}$, given by

$$r((c,e),a) = a\varrho_i - \psi_e(c-a) \quad \text{for } (c,e) \in \mathfrak{C} \times \tilde{\mathfrak{J}}, a \in \mathfrak{A}(d) ,$$

and $r((c,e),a) = 0$ otherwise. Total expected revenue should be maximized, so the discount factor is $\alpha = 1$.

As discussed in Sect. 2.2.2, assumption (A1) implies (AS). Combined with Theorem 2.2, it shows that there is a stationary policy $f^\infty \in \mathfrak{F}^\infty$ that is optimal. Moreover, $V(c,e)$, the maximum expected total reward starting in state (c,e),

$$V(c,e) = \max_{\pi \in \Pi} V_\pi(c,e), \quad (c,e) \in \mathfrak{X} ,$$

is well-defined.

Note that on $\mathfrak{J}_0$, $V \equiv 0$. After having entered the absorbing set, neither additional revenues nor costs incur. For all $f \in \mathfrak{F}$, $(c,e) \in \mathfrak{J}$, $e = (d,i,z)$, all $a \in \mathfrak{A}(d)$, and all $v : \mathfrak{J} \to \mathbb{R}$, we set

$$Lv(c,e,a) = a\varrho_i - \psi_e(c-a) + \sum_{c'=c-a}^{C} q^e_{c-a,c'} \sum_{e' \in \tilde{\mathfrak{J}}} \tilde{p}(e,e')v(c',e') ,$$

$$Uv(c,e) = \max\{Lv(c,e,a) \mid a \in \mathfrak{A}(d)\} .$$

Theorem 2.2 goes on to say that V is the solution to the optimality equation $V = UV$ and may be obtained by value iteration. In addition, every policy consisting of decision rules $f^* \in \mathfrak{F}$ that maximize UV is optimal.

Our objective is to establish the optimality of a decision rule of a (generalized) protection level type. A decision rule $f_y \in \mathfrak{F}$ is called a generalized protection level rule (or of generalized protection level type) if there exists a function $y : \tilde{\mathfrak{J}} \to \mathfrak{A}$ such that for all $c \in \mathfrak{C}$ and $e = (d, i, z) \in \tilde{\mathfrak{J}}$

$$f_y(c, e) = \begin{cases} \min\{d,\ c - y(e)\} & c > y(e) \\ 0 & c \le y(e) \ . \end{cases}$$

Decision rules of the generalized protection level type accept additional requests as long as the remaining capacity is higher than $y(e)$, a number that may depend on the current state of the environmental process. We consider a decision rule to be of protection level type if its protection levels depend solely on the fare class i. If they depend on some additional time parameter, we speak of time-dependent protection levels.

4.2 Structural Results

Theorem 4.1.
Assume (A1) to (A3). There then exists an optimal generalized protection level rule $f_y^* \in \mathfrak{F}$ with protection levels

$$y^*(e) \ge \hat{y}(e) := \sup\{c \in \mathbb{Z} \mid -\Delta\psi_e(c) > \varrho_i\}\ , \quad e \in \tilde{\mathfrak{J}}\ .$$

Proof. First, we prove prove the existence of an optimal generalized protection level rule. Note that we can conclude from assumptions (A2) and (A3)(ii) that an optimal policy overbooks only a finite number of customer requests. In addition, it was shown in Sect. 2.2.2 that (A1) implies (AS). Hence, we can apply Theorem 2.2, which ensures that value iteration works. So let $v_0 \equiv 0$, $v_{n+1} = Uv_n(c, e)$.

We show by induction that the following assertions (i) and (ii) are true for all $n \in \mathbb{N}_0$.

(i) $h_n(c - a, e) = -\psi_e(c - a) + \sum_{c'=c-a}^{C} q^e_{c-a,c'} \sum_{e' \in \tilde{\mathfrak{J}}} \tilde{p}(e, e') v_n(c', e')$ is increasing and concave in c.
(ii) $v_{n+1}(c, e)$ is increasing and concave in c.

At the start of the induction for $n = 0$, (i) is clearly true because of $v_0 \equiv 0$ together with assumptions (A2) and (A3). Thus, $h_0(c - a, e)$ is increasing and concave in c. Using Lemma 1 from Stidham (1978) (see Appendix B), it follows that

$$v_1(c, e) = \max_{a=0,\dots,d} \{a\varrho_i + h_0(c - a, e)\}$$

is increasing and concave in c, i.e. (ii) is true.

Now suppose that (i) and (ii) are true for an integer $n \geq 0$. From (ii), we know that v_{n+1} is increasing and concave. Combining this with assumptions (A2) and (A3) yields (i). Knowing that $h_{n+1}(c-a,e)$ is concave, (ii) is again a simple consequence of Stidham's Lemma 1.

Hence, the induction is complete. It follows that $V(c,e)$ is increasing and concave in c.

Let

$$h(c,e) := -\psi_e(c) + \sum_{c'=c}^{C} q^e_{c,c'} \sum_{e' \in \tilde{\mathfrak{J}}} \tilde{p}(e,e')V(c',e')$$

in order to write

$$V(c,e) = \max_{a=0,\dots,d} \{a\varrho_i + h(c-a,e)\} .$$

Action a is optimal in state (c,e) if

$$LV(c,e,a) = a\varrho_i + h(c-a,e) > a'\varrho_i + h(c-a',e) = LV(c,e,a')$$

holds for $a < a' \leq d$, and

$$LV(c,e,a) = a\varrho_i + h(c-a,e) \geq a'\varrho_i + h(c-a',e) = LV(c,e,a')$$

for $0 \leq a' < a$. Since $h(\cdot,e)$ is increasing and concave, this is equivalent to

$$\begin{aligned} a\varrho_i + h(c-a,e) &> (a+1)\varrho_i + h(c-(a+1),e) \\ a\varrho_i + h(c-a,e) &\geq (a-1)\varrho_i + h(c-(a-1),e) . \end{aligned}$$

These inequalities reduce to only the first (second) of the two inequations in the case of $a = 0$ ($a = d$) and are equivalent to

$$\begin{aligned} \varrho_i &< h(c-a,e) - h(c-a-1,e) \\ \varrho_i &\geq h(c-a+1,e) - h(c-a,e) . \end{aligned}$$

If we define $\sup\{\emptyset\} = \infty$, we can conclude that there is an optimal generalized protection level

$$y^*(e) = \sup\{c \in \mathbb{Z} \mid h(c,e) - h(c-1,e) > \varrho_i\} .$$

Action $a = 0$ is optimal for all $c < y^*(e)$, action $a = d$ for all $c > y(e)^* + d$. For all $y(e)^* \leq c \leq y(e)^* + d$, $a = c - y(e)$ is optimal.

Now let us turn to the lower bounds $\hat{y}(e)$. Due to assumption (A2),

$$\begin{aligned} &h(c,e) - h(c-1,e) \\ &= -\psi_e(c) + \psi_e(c-1) \\ &\quad + \sum_{c'=c}^{C} q^e_{c,c'} \sum_{e' \in \tilde{\mathfrak{J}}} \tilde{p}(e,e')V(c',e') - \sum_{c'=c-1}^{C} q^e_{c-1,c'} \sum_{e \in \tilde{\mathfrak{J}}} \tilde{p}(e,e')V(c',e') \\ &\geq -\psi_e(c) + \psi_e(c-1) = -\Delta\psi_e(c) . \end{aligned}$$

So for the value $\hat{y}(e) = \sup\{c \in \mathbb{Z} \mid -\Delta\psi_e(c) > \varrho_i\}$, it holds that

$$\begin{aligned}\hat{y}(e) &= \sup\{c \in \mathbb{Z} \mid -\Delta\psi_e(c) > \varrho_i\} \\ &\leq \sup\{c \in \mathbb{Z} \mid h(c,e) - h(c-1,e) > \varrho_i\} = y^*(e) ,\end{aligned}$$

and we have a lower bound for $y^*(e)$ (which might be infinity, e.g. if the set is empty). □

It is a direct consequence of Theorem 4.1 that given that there are no costs of unsold capacity, $\psi_e(c) = 0$ for all $c \geq 0$, $y^*(d,0,z) \geq \hat{y}(d,0,z) = \infty$. Seats should not be discarded.

Under the following assumption, it can be shown that the protection levels are independent of the observed demands and that they are increasing in the observed booking class (for any external state).

(A4) $\kappa_e(.) \equiv \kappa_z(.)$, $\psi_e(.) \equiv \psi_z(.)$, and $q^e_{cc'} = q^z_{cc'}$, $c, c' \in \mathfrak{C}$, depend on $e = (d,i,z)$ only through z.

Theorem 4.2. *Assume (A1) to (A4). Then $y^*(e) = y^*(i,z)$ depends on $e = (d,i,z)$ only through i and z and for all booking classes $i \leq i'$, $i \neq 0$, we have $y^*(i,z) \leq y^*(i',z)$.*

Proof. Fix $e = (d,i,z)$ and $e' = (d',i',z')$ with $i \leq i'$, $i \neq 0$, and $z = z'$. For $c \in \mathfrak{C}$ set

$$\begin{aligned}h(c,z) := &-\psi_z(c) \\ &+ \sum_{c''=c}^{C} q^z_{c,c''} \sum_{z'' \in \mathfrak{E}} \kappa_z(z'') \sum_{(d'',i'') \in \mathfrak{D} \times \mathfrak{I}} \eta_{z''}(d'',i'') V(c'',(d'',i'',z'')) ,\end{aligned}$$

so that

$$V(c,e) = \max_{a=0,\dots,d} \{a\varrho_i + h(c-a,z)\} .$$

We know from the proof of Theorem 4.1 that $h(c,z)$ is concave. Therefore, $y^*(e)$ can be characterized as an action $a_0 = \max\{0, c - y^*(e)\}$, say, such that both

$$a\varrho_i + h(c-a,z) \leq (a+1)\varrho_i + h(c-a-1,z) , \quad a < a_0 ,$$

and

$$a\varrho_i + h(c-a,z) \geq (a+1)\varrho_i + h(c-a-1,z) , \quad a \geq a_0 ,$$

hold, from which we infer that $y^*(e)$ is independent of d.

If $y^*(i',z) = \infty$, then $y^*(i,z) \leq y^*(i',z)$ trivially holds. Therefore, let $y^*(i',z) < \infty$. Given $c > y^*(i',z)$,

$$h(c, z) - h(c - 1, z) \le \varrho_{i'} \ .$$

As $\varrho_i \ge \varrho_{i'}$ for $i \ne 0$ by assumption, it follows that for $c > y^*(i', z)$,

$$h(c, z) - h(c - 1, z) \le \varrho_i$$

also holds, which ultimately yields $y^*(i, z) \le y^*(i', z)$ for all $i \ne 0$. □

4.3 Examples for the Random Environment

We have still not answered what the external process (Z_n) could stand for. Thus, in this section, our objective is to demonstrate the great versatility of the environmental process. To avoid technical difficulties and cumbersome notation, we will often restrict our attention to simple situations dealing with only one specific topic. Clearly, most of these features can also be realized simultaneously.

4.3.1 Exogenous Effects Based on the Evolution of a Markov Chain

Assume (Z_n) to be an exogenous Markov chain with state space $\mathfrak{Z}$ and transition matrix $\tilde{P}^{\mathfrak{Z}}$, which is independent of the arrival process of the customers.

This Markov chain might represent the overall economic cycle or the change in currency exchange rates. Such factors might influence the booking behavior of customers with respect to the number of requests at a certain point in time, the booking classes requested, the cancelation probabilities, and, finally, the terminal costs.

Other uncertain effects, like the recovery in tourism in regions that have suffered from natural catastrophes or terrorist attacks, might also be modeled by such an external Markov chain. As a very simple case, one could also consider a Markov chain representing a pure time parameter that counts the number of periods until departure.

We give a numerical example of capacity control with exogenous effects based on the evolution of a Markov chain in the next subsection.

4.3.2 Capacity Control Under Uncertainty

For new routes or flight time changes, there is often some uncertainty about the transition matrix $\tilde{P}$ of the environmental process (E_n). For example, one of the following objects (or combinations thereof) might be only partially known:

(1) the transition matrix $\tilde{P}^{\mathfrak{Z}}$ of the exogenous Markov chain considered above,
(2) the distribution of the demands D_n, provided that they are i.i.d. random variables,

(3) the distribution of the fare classes I_n, provided that they are i.i.d. random variables.

To illustrate the idea of an adaptive control of such a system under uncertainty, let the fare classes $I_0, I_1, \ldots$ be i.i.d. random variables with distribution $P(I_n = i) = \phi_\vartheta(i)$, $i \in \Im$, which is known up to some unknown parameter $\vartheta \in \mathfrak{T}$.

Handling the uncertainty about ϑ from the Bayesian point of view, some probability measure ν on (the Borel σ-algebra of) $\mathfrak{T}$ is supposed to be given. Based on the prior information ν_{n-1} and the observed class i_{n-1}, the update (posterior information) ν_n then gives the actual information about ϑ at stage n.

To avoid technical difficulties, we assume that the set of probability distributions can be approximated by a finite set. This is not very restrictive if e.g. the set of parameters is compact. Then the set of probability distributions is also compact, and it can thus be approximated arbitrarily well by finite sets.

Finally, by considering the external states $z_n = \nu_n$, there is an (approximated Bayes) optimal decision rule, which is of a generalized protection level type with protection levels $y^*(d_n, i_n, \nu_n)$ explicitly depending on ν_n. Note, however, that in order to fulfill (A1), the environmental process has to be augmented e.g. by a time parameter.

4.3.3 General Demand Patterns

The construction of the external process enables us to consider dependencies between the actual demand and the demand observed at earlier stages.

For example, let $z_n = (z_{1,n}, \ldots, z_{i_{\max},n})$ denote the vector of the total demands $z_{1,n}, \ldots, z_{i_{\max},n}$ of the tickets for the fare classes $1, \ldots, i_{\max}$ up to stage $n-1$. Then, by updating z_n to

$$z_{n+1} = (z_{1,n} + d_n \delta_{1,i_n}, \ldots, z_{i_{\max},n} + d_n \delta_{i_{\max},i_n})$$

(with $\delta_{ii'} = 1$ for $i = i'$ and 0, otherwise), general demand patterns can be modeled easily. Again, this environment has to be augmented by a time parameter in order to fulfill (A1). In this case, there is an optimal decision rule that is of a generalized protection level type with protection levels $y^*(d_n, i_n, z_n)$. Another example of capacity control with dependencies between the actual demand and the demand observed at earlier stages will be given in the next subsection.

Note that this is not inconsistent with Chatwin's finding (1998) that policies of protection level type need not be optimal in the presence of stochastically dependent requests. In contrast to our definition of generalized protection levels, Chatwin forbids protection levels to depend on d or z.

4.4 Numerical Examples

We will now present two simple examples of capacity control in a random environment in order to illustrate our results. In the first example, external effects based on the evolution of a Markov chain are included; the second example uses an environment to model dependencies between future demand and demand observed at earlier stages.

4.4.1 Example 1: Capacity Control with Exogenous Effects

Let (Z_n) be an exogenous Markov chain with state space $\mathfrak{Z}$ and transition matrix $\tilde{P}^{\mathfrak{Z}}$, which is independent of the customer arrival process.

This Markov chain might represent e.g. the general popularity of a certain flight destination, which influences customer booking behavior to some degree with respect to the number of requests at a certain point in time and the booking classes requested. In order to fulfill (A1), we also incorporate a time parameter t to count the periods until departure.

We assume that the plane is to depart after $N = 10$ time periods. In every period, a popularity Markov chain with two states $\mathfrak{Z}^\ell = \{1, 2\}$ determines the demand for this flight. Customers consider the destination to be more attractive in state $\ell = 2 \in \mathfrak{Z}^\ell$ than in $\ell = 1 \in \mathfrak{Z}^\ell$.

There are $i_{\max} = 5$ fare classes with $\varrho_5 = 100 < \varrho_4 = 200 < \varrho_3 = 300 < \varrho_2 = 400 < \varrho_1 = 500$. In state $\ell = 2$, there is a probability of only 0.1 for a period without any demand for reservations. With a probability of 9/200 each, there are requests for 1, 2, 3, or 4 reservations in classes 1 to 5 within one period. In state $\ell = 1$, the probability of no demand arrival within a period is 0.2. There is only demand for 1 or 2 reservations in classes 3, 4, or 5 with equal probability. Consequently,

$$\eta_{(\ell_n, t_n)}(d_n, i_n) = \begin{cases} 1/5 & \text{for } \ell_n = 1,\ t_n \neq 0,\ d_n = 1,\ i_n = 0, \\ 2/15 & \text{for } \ell_n = 1,\ t_n \neq 0,\ d_n \in \{1, 2\}, i_n \in \{3, 4, 5\}, \\ 1/10 & \text{for } \ell_n = 2,\ t_n \neq 0,\ d_n = 1,\ i_n = 0, \\ 9/200 & \text{for } \ell_n = 2,\ t_n \neq 0,\ d_n \in \{1, 2, 3, 4\}, \\ & \quad i_n \in \{1, 2, 3, 4, 5\}, \\ 1 & \text{for } t_n = 0,\ d_n = 1,\ i_n = 0, \\ 0 & \text{otherwise.} \end{cases}$$

The external Markov chain has the state space $\mathfrak{Z} = \mathfrak{Z}^\ell \times \{0, 1, \ldots, 10\}$ with each state $z = (l, t)$ describing both the popularity ℓ and the periods until departure t. The popularity state changes from one period to the next with a probability of 0.3; it remains the same otherwise. Before departure, the time parameter t always reduces by one, i.e.

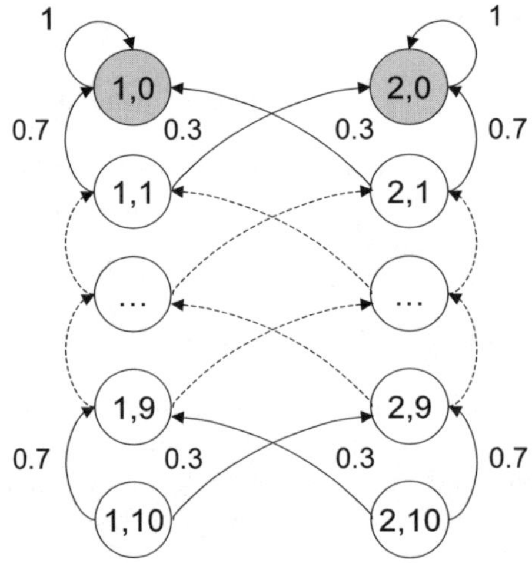

Fig. 4.1. Transition graph of the external Markov chain from Example 1.

$$\kappa'_{(\ell_n,t_n),d_n,i_n}((\ell_{n+1},t_{n+1})) = \begin{cases} 7/10 & \text{for } \ell_n = \ell_{n+1},\ t_{n+1} = t_n - 1, \\ 3/10 & \text{for } \ell_n \neq \ell_{n+1},\ t_{n+1} = t_n - 1, \\ 1 & \text{for } t_{n+1} = t_n = 0, \\ 0 & \text{otherwise.} \end{cases}$$

At time $t = 0$, the plane departs and no more demand arrives. Therefore, the absorbing set is $\tilde{\mathfrak{J}}_0 = \{(d,i,(\ell,t)) \in \mathfrak{Z} \mid t = 0\}$. The transition graph of the external Markov chain is given in Fig. 4.1. Absorbing states are shaded in gray.

We assume that there are a total of $C = 5$ seats to sell and do not allow for no-shows. The number of cancelations is assumed to follow a binomial distribution with a cancelation probability of 0.1. In addition, we discuss the case in which there are no cancelations ($q^e_{c,c'} = 1$ for $c' = c$, $e \in \tilde{\mathfrak{J}}$, and 0 otherwise). In both settings, (A2) is fulfilled. Penalty costs of 750 per seat if we have less than $-C$ remaining seats and terminal costs of 750 per seat in the case of denied boardings, i.e. $\psi^{\mathrm{DB}}_{\zeta}(x) = 750x$, ensure (A3). Hence, we know from Theorem 4.1 that the optimal policy is of the generalized protection level type with protection levels $y^*(e)$ that depend on the state e of the environmental process. In addition, $y^*(d,0,(\ell,t)) = \infty$.

Now let us turn to the values of $\hat{y}(e)$. For all $e = (d,i,(\ell,t))$ where $t > 1$, it is an immediate consequence of the definition of the penalty costs that $\hat{y}(e) = -C$. Therefore, let us consider $t = 1$. Given no cancelations, $\hat{y}(d,i,(\ell,1)) = 0$ for $i \geq 1$ due to the high costs of denied boardings. In the case of cancelations, the expected costs associated with one person overbooked at $t = 1$ are $(1 - 0.9^6)750 \approx 351$, and the expected costs for two persons overbooked are $(1 - 0.9^6)1500 + (1 - 6 \cdot 0.1 \cdot 0.9^5)750 \approx 1187$. Therefore, $\hat{y}(d,i,(\ell,1)) = 0$ for $i > 2$, and $\hat{y}(d,i,(\ell,1)) = -1$ for $i \in \{1,2\}$.

Table 4.1. Protection levels $y^*(d, i, (w, t))$ of an optimal policy without (and including) cancelations.

	$\ell = 1$			$\ell = 2$				
	$i = 5$	$i = 4$	$i = 3$	$i = 5$	$i = 4$	$i = 3$	$i = 2$	$i = 1$
$t = 1$	0 (0)	0 (0)	0 (0)	0 (0)	0 (0)	0 (0)	0 (-1)	0 (-1)
$t = 2$	1 (0)	0 (0)	0 (-1)	3 (0)	1 (0)	0 (-1)	0 (-1)	0 (-2)
$t = 3$	2 (0)	1 (0)	0 (-1)	4 (2)	3 (0)	1 (-1)	0 (-2)	0 (-4)
$t = 4$	4 (1)	2 (0)	0 (-2)	5 (3)	4 (0)	1 (-2)	0 (-4)	0 (-4)
$t = 5$	5 (3)	3 (0)	0 (-2)	5 (5)	5 (0)	2 (-3)	0 (-3)	0 (-3)
$t = 6$	5 (4)	4 (0)	1 (-3)	5 (5)	5 (0)	3 (-4)	0 (-4)	0 (-4)
$t = 7$	5 (5)	5 (-1)	2 (-1)	5 (5)	5 (0)	3 (0)	0 (0)	0 (0)
$t = 8$	5 (5)	5 (-1)	3 (-1)	5 (5)	5 (-1)	4 (-1)	0 (-1)	0 (-1)
$t = 9$	5 (5)	5 (-2)	3 (-2)	5 (5)	5 (-1)	5 (-1)	0 (-1)	0 (-1)
$t = 10$	5 (5)	5 (-2)	4 (-2)	5 (5)	5 (-2)	5 (-2)	1 (-2)	0 (-2)

Since (A4) holds in this example, we can conclude from Theorem 4.2 that the protection levels only depend on the environment $z = (\ell, t)$ and the booking class i. Protection levels of an optimal policy are given in Table 4.1. Assuming that the initial distribution of the popularity Markov chain is $(0.5, 0.5)$, the total expected revenue from the selling process is about 3081 in the case of cancelations and 1797 without. (The revenue figure with cancelations is so much higher because there are no refunds.) Since class 1 customers offer the highest possible revenue per seat, protection levels are 0 for $i = 1$ if there are no cancelations. In accordance with Theorem 4.2, the protection levels are increasing in i, i.e. the values of the protection levels are higher for low-fare demand. If no cancelations are considered, protection levels decrease for decreasing values of t. As expected, protection levels are lower in the case of cancelations and may even be negative. Note that in the case of cancelations, protection levels are not monotone in t. In addition, the protection levels of classes $i = 4$ and 5 given a popularity of $\ell = 2$ are higher than given $\ell = 1$ for all $t = 1, \ldots, 10$. The optimal policy is more discriminating in $\ell = 2$ than in $\ell = 1$, since the probability of future high-value demand is greater.

4.4.2 Example 2: Capacity Control with Dependent Demand

As another example, one could imagine the following: Let us consider an aircraft with a capacity of $C = 10$ seats. Demand arrives in one of $i_{\max} = 3$ fare classes with associated revenues of $\varrho_3 = 100 < \varrho_2 = 200 < \varrho_1 = 300$.

Suppose the external process (Z_n) is a Markov chain with $\mathfrak{Z} = \{0, 1\}$, where state 1 represents (additional) arriving customer demand, and state 0 represents the end of the arrival process.

In state 1, there is demand for 1 or 2 reservations for classes 1, 2, and 3 with a probability of 1/6 each, i.e.

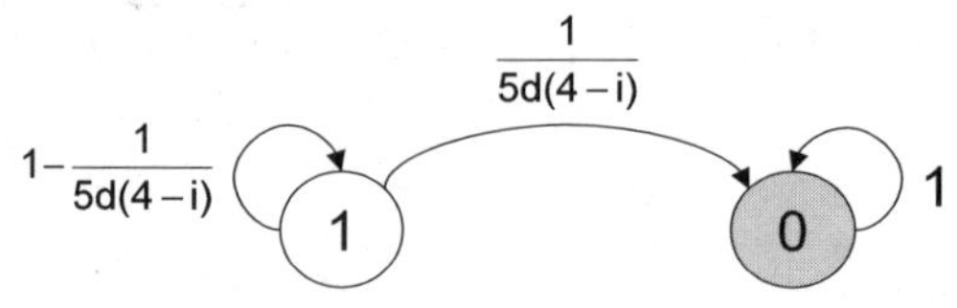

Fig. 4.2. Transition graph of the external Markov chain from Example 2.

$$\eta_{z_n}(d_n, i_n) = \begin{cases} 1/6 \text{ for } z_n = 1,\ d_n \in \{1,2\},\ i_n \in \{1,2,3\}, \\ 1 \quad \text{for } z_n = 0,\ d_n = 1,\ i_n = 0, \\ 0 \quad \text{otherwise.} \end{cases}$$

The probability of more demand arriving before departure depends on the quality and quantity of the current demand. The more demand observed and the higher the associated fare, the more probable the arrival of additional demand before departure is. The probability of more demand arriving before departure given a demand of d from class $i \in \{1, 2, 3\}$ is $1 - 1/(5d(4 - i))$. Therefore, the probability of no additional demand before departure, a transition to state 0, is $1/(5d(4 - i))$.

To sum this up, the transition probabilities of the Markov chain can be stated as

$$\kappa_{z_n,d_n,i_n}(z_{n+1}) = \begin{cases} 1/\left(5d_n(4 - i_n)\right) & \text{for } z_n = 1,\ z_{n+1} = 0,\ d_n \in \{1,2\}, \\ & i_n \in \{1,2,3\}, \\ 1 - 1/\left(5d_n(4 - i_n)\right) & \text{for } z_n = 1,\ z_{n+1} = 1,\ d_n \in \{1,2\}, \\ & i_n \in \{1,2,3\}, \\ 1 & \text{for } z_n = 0,\ z_{n+1} = 0,\ d_n = 1, \\ & i_n = 0, \\ 0 & \text{otherwise.} \end{cases}$$

The corresponding transition graph is given in Fig. 4.2. The absorbing state is shaded in gray.

For simplicity, we assume that there are neither cancelations nor no-shows. There are no penalty costs, i.e. $\psi^{\mathrm{P}}(c, e) = 0$. Terminal costs are 0 if $c \geq 0$ and increase linearly by 120 for every passenger overbooked, i.e. $\psi^{\mathrm{T}}(c, e) = \max\{0, -120c\}$.

First, let us check for the assumptions (A1) to (A4). Since $\tilde{\mathfrak{J}}_0 = \{e = (d, i, z) \in \mathfrak{E} \mid z = 0\}$, it follows that

$$\|\tilde{P}_{\tilde{\mathfrak{J}}} 1\| = 1 - \frac{1}{5d(4 - i)} < 1 \,.$$

Thus, assumption (A1) holds for $m = 1$. (A2) is trivially true. Since

$$\psi_e(c) = \sum_{c'=c}^{C} q^e_{c,c'} \frac{1}{5d(4 - i)} \max\{0, -120c'\} \geq \sum_{c'=c}^{C} q^e_{c,c'} \max\{0, -4c'\} \,,$$

Table 4.2. Protection levels $y^*(d,i,1)$ of an optimal policy.

	$i=3$	$i=2$	$i=1$
$d=1$	7	1	0
$d=2$	9	2	0

assumption (A3) can easily be seen to hold.

We can therefore conclude from Theorem 4.1 that the optimal policy is of generalized protection level type with protection levels $y^*(e)$ that depend on the state e of the environmental process. Since in this example κ_e depends on i and d, however, assumption (A4) is not fulfilled, and the protection levels need not be independent of d.

Protection levels of an optimal policy are given in Table 4.2. Since class 1 customers offer the highest revenue possible per reservation, it is obvious that protection levels are 0 for $i=1$. Since low demand increases the probability for the end of the booking horizon, protection levels are increasing in d.

In this simple example, the protection levels of an optimal policy are increasing in i, too. But this does not need to be true in general. Emphasizing the effect of low value demand as an indicator for the end of the booking horizon by setting e.g.

$$\kappa'_{z_n,d_n,i_n}(z_{n+1}) = \begin{cases} 6/10 & \text{for } z_n=1, z_{n+1}=0, i_n=3, \\ & d_n \in \{1,2\}, \\ 4/10 & \text{for } z_n=1, z_{n+1}=1, i_n=3, \\ & d_n \in \{1,2\}, \\ 1/\left(5d_n(4-i_n)\right) & \text{for } z_n=1, z_{n+1}=0, i_n \in \{1,2\}, \\ & d_n \in \{1,2\}, \\ 1-1/\left(5d_n(4-i_n)\right) & \text{for } z_n=1, z_{n+1}=1, i_n \in \{1,2\}, \\ & d_n \in \{1,2\}, \\ 1 & \text{for } z_n=0, z_{n+1}=0, i_n=0, \\ & d_n=1, \\ 0 & \text{otherwise} \end{cases}$$

and decreasing the probability of low-value demand e.g. to

$$\eta'_{z_n}(d_n,i_n) = \begin{cases} 1/100 & \text{for } z_n=1, i_n=3, d_n \in \{1,2\}, \\ 245/1000 & \text{for } z_n=1, i_n \in \{1,2\}, d_n \in \{1,2\}, \\ 1 & \text{for } z_n=0, i_n=0, d_n=1, \\ 0 & \text{otherwise} \end{cases}$$

leads to the protection levels of an optimal policy as given in Table 4.3. Although the general structure discussed above remains intact, the low probability of additional customer arrivals given the low-value demand causes extremely low protection levels given that $i=3$.

Table 4.3. Protection levels $y^*(d, i, 1)$ of an optimal policy assuming an emphasized effect of low-value demand.

	$i = 3$	$i = 2$	$i = 1$
$d = 1$	2	3	0
$d = 2$	2	4	0

5

Basic Single Resource Capacity Control Models in Revenue Management

The two main textbook models of single-resource capacity control are the dynamic and the static capacity control model. Both models fulfill assumptions i) to xi) mentioned in Sect. 1.1.1; they differ only in the assumptions concerning the arrival process. Static capacity control models assume that demand for the different booking classes arrives in non-overlapping periods. Dynamic capacity control models allow passengers to arrive in any order. In turn, they assume demand to be Markovian.

Most capacity control models can be assigned to one of these two categories. One exception is the so-called omnibus model, which generalizes the two models by allowing more than one customer of a certain booking class to arrive in the framework of a dynamic model. It was proposed by Lautenbacher and Stidham (1999) and Mayer (1976). Another exception is a semi-Markov decision model described in Walczak (2001) and Brumelle and Walczak (2003).

5.1 The Dynamic Model

In the basic dynamic model, the booking horizon $[0, T]$ is divided into N time periods so that the probability of two or more requests arriving within one period can be neglected. These periods are indexed by n and the indices run backwards in time. Consequently, smaller values of n indicate later points in time. Period N corresponds to the first period within the booking horizon, while period 1 denotes the last period with positive arrival probability. Thus, period 0 can be interpreted as the scheduled departure time. For every period $n = 1, \ldots, N$, the probability of a class i customer request is given by $\hat{p}_n(i)$. As before, fare classes are ordered such that $0 < \varrho_{i_{\max}} < \varrho_{i_{\max}-1} < \cdots < \varrho_1$. Furthermore, $\hat{p}_n(0) = 1 - \sum_{i=1}^{i_{\max}} \hat{p}_n(i)$ denotes the probability of no customer request in period n. Neither cancelations nor no-shows are allowed. Demand is independent of the controls used.

In this setting, dynamic models answer the question of whether or not to accept a particular reservation request for booking class i in period n given a remaining capacity of c seats.

A dynamic capacity control model with only two fare classes was introduced by Gerchak et al. (1985). Lee and Hersh (1993) extended their model to allow for multiple fare classes. Lautenbacher and Stidham (1999) stressed the structure of the underlying Markov decision process of the dynamic model. Liang (1999) considers the continuous time case.

Extensions of this model include a two-class model with customer choice (i.e. without assumption x)) described in Weatherford et al. (1993). In another two-class model with customer choice, Zhao and Zheng (2001) assume that the discount fare cannot be reopened. Similarly, Feng and Xiao (2000) decide on when to stop offering the low-fare class when a certain number of capacity units is reserved for high-fare customers in order to ensure a minimum service level for that price segment. Lee and Hersh (1993) allow for group arrivals. No-shows, cancelations, overbooking, and refunding are considered in Subramanian et al. (1999), Talluri and van Ryzin (2004b, pp. 155–160), and Feng and Xiao (2006a).

5.1.1 The Decision Model

The objective of finding a policy that maximizes expected revenue in the basic dynamic model is usually reduced to solving the optimality equation of a finite-stage Markov decision model $(N, \mathfrak{X}, \mathfrak{A}_n, p_n, r_n, V_0)$ with planning horizon N. The state space is $\mathfrak{X} = \mathfrak{C} \times \mathfrak{I} = \{(c,i) \in \mathbb{Z} \times \mathbb{N}_0 \mid c \leq C,\ i \leq i_{\max}\}$, in which we refer to c as the remaining capacity and to i as the requested booking class; $i = 0$ denotes the artificial class 0 with fare $\varrho_0 = 0$. The action space, $\mathfrak{A} = \mathfrak{A}_n = \{0,1\} \equiv \{\text{reject}, \text{accept}\}$ for all n, specifies the sets of admissible actions in state $(c,i) \in \mathfrak{X}$, $\mathfrak{A}(c,i) = \mathfrak{A}$ for $i = 1, \ldots, i_{\max}$ and $\mathfrak{A}(c,0) = \{0\}$. The transition laws p_n from $\mathfrak{K}_n = \{(c,i,a) \mid (c,i) \in \mathfrak{X}, a \in \mathfrak{A}(c,i)\}$ into $\mathfrak{X}$, for $n = N, N-1, \ldots, 1$ are defined by $p_n((c,i), a, (c-a, i')) = \hat{p}_{n-1}(i')$ and 0 otherwise. The one-stage reward functions r_n on $\mathfrak{K}_n$ are given by $r_n((c,i),a) = a\varrho_i$, and the terminal reward function V_0 on $\mathfrak{X}$ by $V_0(c,i) = 0$ for $c \geq 0$ and $V_0(c,i) = c\bar{\varrho}$ for $c < 0$ with $\bar{\varrho} > \varrho_1$.

By $(X_N, X_{N-1}, \ldots, X_0)$, we denote the state process of the MDP, and by $\mathfrak{F}^N$ the set of all policies. We write $R_\pi = \sum_{n=1}^N r_n(X_n, f_n(X_n)) + V_0(X_0))$ for the total reward gained when applying policy π. From Theorem 2.1, we know that for $(c,i) \in \mathfrak{X}$,

$$\begin{aligned} V^*(c,i) &= \max_{\pi \in \mathfrak{F}^N} \mathbb{E}_\pi \left[\sum_{n=1}^N r_n(X_n, f_n(X_n)) + V_0(X_0)) \mid X_N = (c,i) \right] \\ &= \max_{\pi \in \mathfrak{F}^N} \mathbb{E}_\pi \left[R_\pi \mid X_N = (c,i) \right] , \end{aligned}$$

the maximum expected revenue starting with capacity c given a request from class i, is the unique solution $V_N \equiv V^*$ to the optimality equation

$$V_n(c,i) = \max_{a \in \mathfrak{A}(c,i)} \left\{ a\varrho_i + \sum_{i'=0}^{i_{\max}} \hat{p}_{n-1}(i') V_{n-1}(c-a,i') \right\}, \tag{5.1}$$

which can be obtained for $n = 1, \ldots, N$ iteratively, starting with V_0. Moreover, every policy $\pi^* \in \mathfrak{F}^N$ formed by actions $a = f_n^*(c,i)$, each maximizing the right-hand side of (5.1) is optimal, i.e. leads to V^*.

Using $\varrho_0 = 0$ and the fact that the value function is increasing in c (see Lemma 5.1), one can conclude that

$$V_n(c,0) = A_{n-1}V_{n-1}(c) \geq A_{n-1}V_{n-1}(c-1), \quad n = 1, \ldots, N, c \in \mathfrak{C},$$

where $A_n V_n(c)$ is written in place of $\sum_{i=0}^{i_{\max}} \hat{p}_n(i) V_n(c,i)$. This allows us to extend $\mathfrak{A}(c,0)$ to $\mathfrak{A}$ without loss of generality and write

$$\begin{aligned} V_n(c,i) &= \max_{a \in \{0,1\}} \left\{ a\varrho_i + \sum_{i'=0}^{i_{\max}} \hat{p}_{n-1}(i') V_{n-1}(c-a,i') \right\} \\ &= \max_{a \in \{0,1\}} \left\{ a\varrho_i + A_{n-1}V_{n-1}(c-a) \right\} \end{aligned} \tag{5.2}$$

instead of (5.1).

In order to reduce the complexity of computation, the optimality equation is frequently also stated as

$$A_n V_n(c) = \sum_{i=0}^{i_{\max}} \hat{p}_n(i) \max_{a \in \{0,1\}} \left\{ a\varrho_i + A_{n-1}V_{n-1}(c-a) \right\}, \tag{5.3}$$

with $A_0 V_0(c) = \sum_{i=0}^{i_{\max}} \hat{p}_n(i) V_0(c,i)$; see Sect. 2.1. This reduction of the state space is crucial in applications. But since we are interested in an analysis of the structure of an optimal policy, the two forms (5.2) and (5.3) are equivalent to us.

5.1.2 Structural Results

For the dynamic model, Lee and Hersh (1993) and Lautenbacher and Stidham (1999) prove structural results of an optimal policy; an overview of the results is given in Talluri and van Ryzin (2004b, pp. 76–79).

Using Lemma 1 from Stidham (1978) (see Appendix B), one can show by induction on n that the value function is increasing and concave in c; see e.g. Lautenbacher and Stidham (1999).

Lemma 5.1. *For $n = 1, \ldots, N$, $i = 0, \ldots, i_{max}$ the value function $V_n(c,i)$ is increasing and concave in c.*

In addition, it is easy to see the optimality of accepting an arbitrary request if the number of remaining seats is larger than the maximum number of future requests. A request is rejected in cases where $c \leq 0$.

Lemma 5.2. *For all $n \in \{1, \dots, N\}$, and $i \in \{1, \dots, i_{max}\}$ we have*

(i) $V_n(c, i) = \varrho_i + A_{n-1}V_{n-1}(c-1) = \varrho_i + \sum_{n'=1}^{n-1} \sum_{i'=0}^{i_{max}} \hat{p}_{n'}(i')\varrho_{i'}, \quad c \geq n.$
(ii) $V_n(c, i) = A_{n-1}V_{n-1}(c) = \bar{\varrho}c, \quad c \leq 0.$

Proof. (i) and (ii) follow by induction on n. Consider $n = 1$. For $c \geq 1$, the terminal reward is $V_0(c, 0) = 0$ by assumption. Therefore, (i) directly follows from

$$V_1(c, i) = \max\{\varrho_i, 0\} = \varrho_i \ .$$

For $c \leq 0$, the terminal reward is $V_0(c, 0) = c\bar{\varrho}$. Making use of $\bar{\varrho} > \varrho_1$ we obtain

$$V_1(c, i) = \max\{c\bar{\varrho} + (\varrho_i - \bar{\varrho}), c\bar{\varrho}\} = c\bar{\varrho} \ ,$$

showing (ii) for $n = 1$.

Now assume (i) and (ii) are true for some $1 \leq n \leq N - 1$. It then follows for $c \geq n + 1$ that

$$\begin{aligned} V_{n+1}(c, i) &= \max\left\{\varrho_i + \sum_{n'=1}^{n} \sum_{i'=0}^{i_{\max}} \hat{p}_{n'}(i')\varrho_{i'}, \sum_{n'=1}^{n} \sum_{i'=0}^{i_{\max}} \hat{p}_{n'}(i')\varrho_{i'}\right\} \\ &= \varrho_i + \sum_{n'=1}^{n} \sum_{i'=0}^{i_{\max}} \hat{p}_{n'}(i')\varrho_{i'} \ . \end{aligned}$$

The optimality equation for $c \leq 0$ at stage $n + 1$ reduces to

$$V_{n+1}(c, i) = \max\{\bar{\varrho}c + (\varrho_i - \bar{\varrho}), \bar{\varrho}c\} = \bar{\varrho}c \ ,$$

which completes the proof. □

Combining these lemmas yields that there is an optimal policy of time-dependent protection level type.

Theorem 5.1. *For the dynamic problem, there exists an optimal policy $\pi^* = (f_N^*, f_{N-1}^*, \dots, f_1^*)$ such that*

$$f_n^*(c, i) = \begin{cases} 1 & c > y_{i-1}^*(n) \\ 0 & c \leq y_{i-1}^*(n) \ , \end{cases}$$

with time-dependent protection levels

$$y_{i-1}^*(n) = \max\{c \in \{0, \dots, n-1\} : \varrho_i < \Delta A_{n-1}V_{n-1}(c)\} \ .$$

The theorem implies that, given a request from customer class i, $i = 1, \dots, i_{\max}$, a number of $y_{i-1}^*(n)$ seats (the so-called time-dependent protection level) is reserved for demand in periods $n-1, \dots, 1$. An incoming request is accepted if and only if at least one seat is not reserved for future demand.

A proof of this theorem can be found e.g. in Lautenbacher and Stidham (1999) or in Lee and Hersh (1993) if their results are combined with Lemma 5.2. Since the dynamic model constitutes a special case of the capacity control model in a random environment, the fact that the decision rules are of time-dependent protection level type (and that the protection levels are increasing in i) follows immediately from Theorems 4.1 and 4.2.

To show that this model is equivalent to a special case of the capacity control model in a random environment introduced in Chap. 4, choose the external state to be a pure time parameter $\mathfrak{Z} = \{0, 1, \ldots, N\}$, counting the number of periods until flight departure at $z = 0$. Accordingly, let $\kappa_e(z') = \kappa_z(z') = 1$ for $z' = \max\{0, \ z-1\}$ and 0 otherwise. The demand D_n, i.e. the number of reservations requested at time n, is equal to 1 almost surely. The probability of a request from booking class i is $\eta_z(1, i) = \hat{p}_z(i)$, and 0 otherwise. The absorbing set is $\tilde{\mathfrak{J}}_0 = \{(d, i, z) \in \mathfrak{E} \mid z = 0\}$. Since cancelations are not allowed, we have $q^e_{c,c'} = 1$ for $c' = c$ and 0 otherwise for all $c \in \mathfrak{C}$, $e \in \mathfrak{E}$. Choosing $\psi^{\mathrm{P}}(c, e) = \bar{\varrho} \max\{0, \ -c\}$, $(c, e) \in \mathfrak{J}_0$, ensures that there is no overbooking. There are no terminal costs.

It easily follows that $\|\tilde{P}^N_{\tilde{\mathfrak{J}}} 1\| = 0$. Therefore, assumption (A1) holds. (A2) holds trivially, since $q^e_{c,c'} = 1$ for $c = c'$ by assumption. Since $\psi_z(c) = \psi^{\mathrm{P}}(c, e) = \bar{\varrho} \max\{0, \ -c\}$ is decreasing and convex, assumption (A3) is fulfilled. (A4) is fulfilled, since all transition probabilities are independent of the current booking class requested, and $d = 1$. Note that in the notation of the dynamic model, $y^*(i, n) = y^*_{i-1}(n)$.

From Theorem 4.2 (and directly from the formula for the protection level in Theorem 5.1), we can conclude that the protection levels increase in booking class i. The non-negativity of the protection levels is a direct consequence of choosing $\bar{\varrho} > \varrho_1$.

Proposition 5.1. *Given a fixed period n before departure, $y^*_{i-1}(n)$ is increasing in i for $i > 0$, i.e.*

$$y^*_{i_{max}-1}(n) > y^*_{i_{max}-2}(n) > \cdots > y^*_0(n) = 0 \; .$$

Another well-known property of the protection levels is that given a fixed booking class, the values of $y^*_{i-1}(n)$ are increasing in n. For a proof, see Lee and Hersh (1993). However, the behavior of $y^*_{i-1}(n)$ over time can be characterized more precisely, as the following proposition shows.

Proposition 5.2. *For $n = 2, \ldots, N$, and $i = 0, \ldots, i_{max}$*

$$y^*_{i-1}(n-1) \leq y^*_{i-1}(n) \leq y^*_{i-1}(n-1) + 1 \; .$$

Proof. Introduce

$$\begin{aligned}\Delta A_n V_n(c) &= A_n V_n(c) - A_n V_n(c-1)\\ &= \sum_{i=0}^{i_{\max}} \hat{p}_n(i)\,[\max\{A_{n-1}V_{n-1}(c), \varrho_i + A_{n-1}V_{n-1}(c-1)\}\\ &\quad - \max\{A_{n-1}V_{n-1}(c-1), \varrho_i + A_{n-1}V_{n-1}(c-2)\}\,]\ .\end{aligned}$$

We know from Lemma 5.1 that $\Delta A_n V_n(c)$ is decreasing in c. Rearranging terms yields on the one hand

$$\begin{aligned}\Delta A_n V_n(c) &= \sum_{i=0}^{i_{\max}} \hat{p}_n(i)\,[\max\{\Delta A_{n-1}V_{n-1}(c), \varrho_i\} + A_{n-1}V_{n-1}(c-1)\\ &\quad - \max\{\Delta A_{n-1}V_{n-1}(c-1), \varrho_i\} - A_{n-1}V_{n-1}(c-2)\,]\\ &= \Delta A_{n-1}V_{n-1}(c-1) + \sum_{i=0}^{i_{\max}} \hat{p}_n(i)\,[\max\{\Delta A_{n-1}V_{n-1}(c), \varrho_i\}\\ &\quad - \max\{\Delta A_{n-1}V_{n-1}(c-1), \varrho_i\}]\\ &\le \Delta A_{n-1}V_{n-1}(c-1)\ ,\end{aligned}$$

since $\Delta A_{n-1}V_{n-1}(c) \le \Delta A_{n-1}V_{n-1}(c-1)$. On the other hand, one can obtain

$$\begin{aligned}\Delta A_n V_n(c) &= \sum_{i=0}^{i_{\max}} \hat{p}_n(i)\,[\max\{0, \varrho_i - \Delta A_{n-1}V_{n-1}(c)\} + A_{n-1}V_{n-1}(c)\\ &\quad - \max\{0, \varrho_i - \Delta A_{n-1}V_{n-1}(c-1)\} - A_{n-1}V_{n-1}(c-1)\,]\\ &= \Delta A_{n-1}V_{n-1}(c) + \sum_{i=0}^{i_{\max}} \hat{p}_n(i)\,[\max\{0, \varrho_i - \Delta A_{n-1}V_{n-1}(c)\}\\ &\quad - \max\{0, \varrho_i - \Delta A_{n-1}V_{n-1}(c-1)\}]\\ &\ge \Delta A_{n-1}V_{n-1}(c)\ .\end{aligned}$$

From these two inequalities, we can conclude that

$$\Delta A_{n-1}V_{n-1}(c) \le \Delta A_n V_n(c) \le \Delta A_{n-1}V_{n-1}(c-1)\ . \tag{5.4}$$

Now, by definition of the protection levels $y_{i-1}^*(n)$, it holds that

$$\varrho_i < \Delta A_n V_n(y_{i-1}^*(n))\ .$$

We can therefore conclude from 5.4 for $c = y_{i-1}^*(n)$ that

$$\varrho_i < \Delta A_n V_n(y_{i-1}^*(n)) \le \Delta A_{n-1}V_{n-1}(y_{i-1}^*(n) - 1)\ ,$$

which implies $y_{i-1}^*(n-1) \ge y_{i-1}^*(n) - 1$. For $c = y_{i-1}^*(n-1)$ we get

$$\varrho_i < \Delta A_{n-1}V_{n-1}(y_{i-1}^*(n-1)) \le \Delta A_n V_n(y_{i-1}^*(n-1))\ ,$$

which implies $y_{i-1}^*(n) \ge y_{i-1}^*(n-1)$ and completes the proof. □

In summary, the structure of an optimal policy is as follows: Given a request from customer class $i = 1, \ldots, i_{\max}$, a non-negative number of $y^*_{i-1}(n)$ seats (the so-called time-dependent protection level of class $i-1$) is reserved for demand in periods $n-1, \ldots, 1$. The protection levels are lower for higher value demand and decrease, by not more than 1 per period, as departure approaches.

5.1.3 A Numerical Example

To illustrate the structural results, we take up an example given by Lee and Hersh (1993): They consider four booking classes with fares $\varrho_1 = 200$, $\varrho_2 = 150$, $\varrho_3 = 120$, and $\varrho_4 = 80$. The capacity of the airplane is $C = 10$, and there are $N = 30$ booking periods; the request probabilities are listed in Table 5.1.

Figure 5.1 shows the time-dependent protection levels $y^*_{i-1}(n)$ of an optimal policy, $y^*_0(n) = 0$. Recall that $n = 0$ corresponds to the flight departure.

Table 5.1. Request probabilities $\hat{p}_n(i)$.

	n				
i	$1 \le n \le 4$	$5 \le n \le 11$	$12 \le n \le 18$	$19 \le n \le 25$	$26 \le n \le 30$
1	0.15	0.14	0.10	0.06	0.08
2	0.15	0.14	0.10	0.06	0.08
3	0	0.16	0.10	0.14	0.14
4	0	0.16	0.10	0.14	0.14

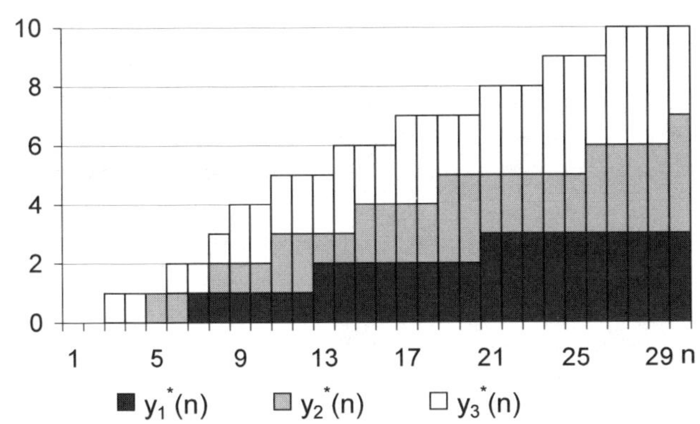

Fig. 5.1. Protection levels of an optimal policy in the case of a risk-neutral decision-maker.

As an example, consider the situation 17 periods before departure. If a class 4 request arrives, it is only accepted given a remaining capacity of 8 or more, since $y_3^*(17) = 7$ seats are protected. If a class 3 request arrives in this period, its acceptance is contingent upon a remaining capacity of 5 or more, since $y_2^*(17) = 4$. Class 2 requests are only rejected if there are $y_1^*(17) = 2$ or fewer seats unsold in this period. Since class 1 customers yield the highest possible revenue per seat, they are always accepted as $y_0^*(17) = 0$.

In accord with Theorem 5.1 and Propositions 5.1 and 5.2, the protection levels are increasing in i and n by increments that do not exceed a height of 1. The expected revenue that is gained from these protection levels is 1403.2.

5.2 The Static Model

In static capacity control models, the demand of each booking class $i \in \{1, \dots, i_{\max}\}$ is supposed to arrive during a single contiguous time segment. In this case, the booking period can be divided into periods with booking requests belonging to the same fare class. At the time the total number of requests, i.e. the demand, d of a booking class i is known, one has to determine the number $a \in \{0, \dots, d\}$ of requests to be accepted in order to maximize the expected revenue of that flight.

The total demands $D_1, \dots, D_{i_{\max}}$ of the booking classes $i = 1, \dots, i_{\max}$ are assumed to be independent random variables on $\mathfrak{D} = \{0, \dots, d_{\max}\}$ with counting densities $P(D_i = d) = \hat{p}_i(d)$, $d \in \mathbb{N}_0$, say. Additionally, it is often assumed that customer requests for tickets arrive in increasing fare order, i.e. the class willing to pay fare $\varrho_{i_{\max}}$ before $\varrho_{i_{\max}-1}$, etc. . We stick to this assumption in the following. Since in this case there is a one-to-one correspondence between periods and classes, we index both by i.

This model is called the "static model", even though requests arrive sequentially over time in stages ordered by booking class and decisions have to be made at every stage. Talluri and van Ryzin (2004b, p. 33) point out that "static" can therefore be seen as a misnomer. Nevertheless, we will use the term in order to adhere to standard terminology and to distinguish our model from the dynamic version (in which the order of incoming requests is arbitrary).

In the common version of the static model, overbooking is not allowed, and cancelations as well as no-shows are not considered. Demand is independent of the controls used.

The static revenue management model was first introduced and solved by Littlewood (1972) for two fare classes. Bhatia and Parekh (1973) and Richter (1982) provide a derivation of the solution, which is extended heuristically to more fare classes by Belobaba (1987a), (1987b), (1989), and Belobaba and Weatherford (1996). Curry (1990), Wollmer (1992), Brumelle and McGill (1993), and Robinson (1995) provide an exact solution to the case of $i_{\max} \geq 2$ fare classes. Li and Oum (2002) unify the optimality conditions given by Curry

(1990), Wollmer (1992), and Brumelle and McGill (1993). Lautenbacher and Stidham (1999) stress the structure of the underlying Markov decision process of the static capacity control model and show structural similarities to the dynamic model.

Extensions of this basic model include a two-class model with dependent demands by Pfeifer (1989), McGill (1989), and Brumelle et al. (1990). Sen and Zhang (1999) determine the optimal initial capacity in this case. Bodily and Weatherford (1995) extend the two-class solution to a heuristic for more than two fare classes. Robinson (1995) relaxes the increasing fare assumption. Bitran and Gilbert (1996) introduce a variant of a three-class model with no-shows for the hotel business. Van Ryzin and McGill (2000) introduce protection levels that adapt their values according to past booking data. Cooper and Gupta (2006) apply stochastic order relations to compare expected revenue gained from optimal protection levels in two markets with the same fare classes but different demand distributions.

Note that although the simplification that customers arrive in increasing fare order seems very restrictive at first sight, it is the model most widely implemented in practice. First, it is argued that discount fares are frequently offered in combination with an advance purchase restriction of several weeks. Customers arriving late have to pay the full price. Second, if customers arrived in a high-to-low fare order, there would never be any reason for rejecting a customer request because no better alternative would become available. From that perspective, the low-to-high fare assumption is similar to a worst-case scenario that must be controlled. Third, due to the aggregated view on booking classes, static models are relatively simple and demand less data than dynamic models. In addition, simulation studies with more realistic passenger behavior by Mayer (1976) and Titze and Griesshaber (1983) show that the results of the static model are still close to optimal in the two-class case.

For a comprehensive introduction to static capacity control models, see Talluri and van Ryzin (2004b, pp. 33–50) or Phillips (2005, pp. 149–174).

5.2.1 The Decision Model

The objective of finding a policy maximizing expected revenue in the static capacity control model is generally reduced to solving the optimality equation of a finite-stage Markov decision model $(i_{\max}, \mathfrak{X}, \mathfrak{A}_i, p_i, r_i, V_0)$ with planning horizon $i_{\max}$. The state space is $\mathfrak{X} = \{(c,d) \in \mathbb{Z} \times \mathfrak{D} \mid c \leq C\}$, where we refer to c as the remaining capacity and to d as the number of requests, i.e. demand, observed for the current booking class. The action space is $\mathfrak{A} = \mathfrak{A}_i = \{0, \dots, d_{\max}\}$ for all i. Action a denotes the number of requests to be accepted. The number of requests that can be accepted is only limited by the observed demand; therefore, the sets of admissible actions in state $(c,d) \in \mathfrak{X}$ is $\mathfrak{A}(c,d) = \{0, \dots, d\}$. The transition laws p_i from $\mathfrak{K}_i = \{(c,d,a) \mid (c,d) \in \mathfrak{X}, a \in \mathfrak{A}(c,d)\}$ into $\mathfrak{X}$ are defined by $p_i((c,d), a, (c-a, d')) = \hat{p}_{i-1}(d')$ and 0 otherwise (with $\hat{p}_0(d)$ arbitrary). The one-stage reward functions r_i on $\mathfrak{K}_i$,

$r_i((c,d),a) = a\varrho_i$ (with ϱ_0 arbitrary), and terminal reward function V_0 on $\mathfrak{X}$, $V_0((c,d)) = 0$ for $c \geq 0$ and $V_0((c,d)) = \bar{\varrho}c$ for $c < 0$ with $\bar{\varrho} > \varrho_1$.

Thus, for booking classes $i = i_{\max}, i_{\max} - 1, \ldots, 1$, given the residual capacity c_i and demand d_i, the number $a_i = f_i(c_i, d_i) \in \{0, \ldots, d_i\}$ of seats to be accepted must be determined.

By $(X_{i_{\max}}, X_{i_{\max}-1}, \ldots, X_0)$, we denote the state process of the MDP, by $\mathfrak{F}^{i_{\max}}$ the set of all policies, and we write $R_\pi = \sum_{i=1}^{i_{\max}} r_i(X_i, f_i(X_i)) + V_0(X_0)$ for the total reward gained when applying policy π. From Theorem 2.1, we know that for $(c,d) \in \mathfrak{X}$,

$$\begin{aligned} V^*(c,d) &= \max_{\pi \in \mathfrak{F}^{i_{\max}}} \mathbb{E}_\pi \left[\sum_{i=1}^{i_{\max}} r_i(X_i, f_i(X_i)) + V_0(X_0) \mid X_{i_{\max}} = (c,d) \right] \\ &= \max_{\pi \in \mathfrak{F}^{i_{\max}}} \mathbb{E}_\pi \left[R_\pi \mid X_{i_{\max}} = (c,d) \right] , \end{aligned}$$

the maximum expected revenue starting with capacity c given d requests from class $i_{\max}$, is the unique solution $V_{i_{\max}} \equiv V^*$ to the optimality equation

$$\begin{aligned} V_i(c,d) &= \max_{a \in \{0,\ldots,d\}} \left\{ a\varrho_i + \sum_{d'=0}^{d_{\max}} \hat{p}_{i-1}(d') V_{i-1}(c-a, d') \right\} \\ &= \max_{a \in \{0,\ldots,d\}} \left\{ a\varrho_i + A_{i-1} V_{i-1}(c-a) \right\} , \end{aligned} \tag{5.5}$$

with $A_i V_i(c) = \sum_{d=0}^{d_{\max}} \hat{p}_i(d) V_i(c,d)$. The solution can be obtained for $i = 1, \ldots, i_{\max}$ by backward induction starting with V_0. Every policy $\pi^* \in \mathfrak{F}^{i_{\max}}$ formed by actions $a^* = f_i^*(c,d)$, each maximizing the right-hand side of (5.5), is optimal.

Again, the optimality equation is frequently referred to as

$$A_i V_i(c) = \sum_{d=0}^{d_{\max}} p_i(d) \max_{a \in \{0,\ldots,d\}} \left\{ a\varrho_i + A_{i-1} V_{i-1}(c-a) \right\} , \tag{5.6}$$

with initial value $A_0 V_0(c) = \sum_{d=0}^{d_{\max}} \hat{p}_i(d) V_0(c,d)$; see e.g. Lautenbacher and Stidham (1999). Due to the reduction of the state space, the complexity of computation is lessened. But since we are mainly interested in structural results of an optimal policy, (5.5) and (5.6) can be used interchangeably.

5.2.2 Structural Results

As in the dynamic model, one can show by induction on n that the value function is increasing and concave in c; see e.g. Lautenbacher and Stidham (1999).

Lemma 5.3. *For $n = 1, \ldots, N$, $d = 0, \ldots, d_{max}$ the value function $V_i(c,d)$ is increasing and concave in c.*

Furthermore, we can show that the counterpart of Lemma 5.2 is the following.

Lemma 5.4. *For* $i \in \{1, \dots, i_{max}\}$, *and* $d \in \{0, \dots, d_{max}\}$ *we have*

(i) $V_i(c,d) = \varrho_i d + A_{i-1}V_{i-1}(c-1) = \varrho_i d + \sum_{i'=1}^{i-1}\sum_{d'=0}^{d_{max}} \hat{p}_{i'}(d')\varrho_{i'}d'$, $c \geq i d_{max}$.

(ii) $V_i(c,d) = A_{n-i}V_{i-1}(c) = \bar{\varrho}c, \quad c \leq 0.$

Proof. Both assertions follow by induction on i. For $n = 1$, the value function is

$$V_1(c,d) = \max_{a=0,\dots,d} \{\varrho_1 a + V_0(c-a,0)\} .$$

Using $V_0(c-a,0) = (c-a)\bar{\varrho}$ for $c \leq 0$, it follows that

$$\varrho_1 a + V_0(c-a,0) = \bar{\varrho}c + a(\varrho_1 - \bar{\varrho}) ,$$

where the last term decreases in a, since $\bar{\varrho} > \varrho_1$. Hence, action $a = 0$ maximizes the value function fulfilling (ii). For $c \geq d_{\max}$, every action results in a non-negative capacity at $i = 0$ with a terminal reward of 0. Consequently, the only term of interest is $\varrho_1 a$, which is increasing in a. The maximum is attained for $a = d$ and (i) is shown.

Now assume that (i) and (ii) are true for some i. Using the induction hypothesis for $c \leq 0$, $V_{i+1}(c,d)$ reduces again to

$$V_{i+1}(c,d) = \max_{a=0,\dots,d} \{\bar{\varrho}c + a(\varrho_{i+1} - \bar{\varrho})\} ,$$

where the maximum is attained for $a = 0$ showing (ii). For $c \geq (i+1)d_{\max}$, every action causes a transition to a state $c - a \geq i d_{\max}$. Using the induction hypothesis, $V_{i+1}(c,d)$ turns to

$$V_{i+1}(c,d) = \max_{a=0,\dots,d} \left\{ \varrho_{i+1}a + \sum_{i'=1}^{i}\sum_{d'=0}^{d_{\max}} \hat{p}_{i'}(d')\varrho_{i'}d' \right\} .$$

Again, $a = d$ maximizes the right-hand side, showing (i). □

From these lemmas, we can conclude that there is an optimal policy of protection level type.

Theorem 5.2. *For the static problem there exists an optimal policy* $\pi^* = (f^*_{i_{max}}, f^*_{i_{max}-1}, \dots, f^*_1)$ *such that*

$$f^*_i(c,d) = \begin{cases} \min\{d, c - y^*_{i-1}\} & c > y^*_{i-1} \\ 0 & c \leq y^*_{i-1} , \end{cases}$$

with protection levels

$$y^*_{i-1} = \max\{c \in \{0, \dots, (i-1)d_{max}\} : \varrho_i < \Delta A_{i-1}V_{i-1}(c)\} .$$

Hence, y^*_{i-1} seats are reserved (protected) for future demand that pays a higher fare. Incoming requests are accepted as long as there are non-protected seats.

Given arrivals in increasing fare order, these protection levels are increasing in i.

Proposition 5.3. *Optimal protection levels are increasing in i, i.e. $y^*_{i_{max}-1} \geq y^*_{i_{max}-2} \geq \cdots \geq y^*_1 \geq y^*_0 = 0$.*

For a complete proof of the structure of an optimal policy, see e.g. Wollmer (1992), Lautenbacher and Stidham (1999), or Talluri and van Ryzin (2004b, pp. 58–62) in combination with Lemma 5.4. Since the static model is a special case of the capacity control model in a random environment, however, the fact that the decision rules are of protection level type and independent of d follows immediately from Theorems 4.1 and 4.2.

To see that this is true, choose the external state to be a pure time parameter $\mathfrak{Z} = \{0, 1, \ldots, i_{\max}\}$ and let $\kappa_e(z') = \kappa_z(z') = 1$ for $z' = \max\{0, \; z-1\}$ and 0 otherwise. The booking class I_n is almost surely equal to z. Hence, $\eta_z(d,i) = \hat{p}_z(d)$, and the absorbing set is $\tilde{\mathfrak{Z}}_0 = \{(d,i,z) \in \mathfrak{E} \mid z = 0\}$. Since cancelations are not allowed, we have $q^e_{c,c'} = 1$ for $c' = c$ and 0 otherwise ($c \in \mathfrak{C}$, $e \in \mathfrak{E}$). Let $\psi^{\mathrm{P}}(c,e) = \bar{\varrho}\max\{0, \; -c\}$, $(c,e) \in \mathfrak{Z}_0$. There are no terminal costs.

Again, $\|\tilde{P}^{i_{\max}}_{\tilde{\mathfrak{Z}}} 1\| = 0$ and assumption (A1) is fulfilled. Since there are no cancelations, $q^e_{c,c'} = 1$ for $c = c'$ and (A2) holds. Again, $\psi_z(c) = \psi^{\mathrm{P}}(c,e) = \bar{\varrho}\max\{0, \; -c\}$ is decreasing and convex by assumption. Thus, (A3) is fulfilled. Since the transition probabilities do not depend on the realization of d, (A4) holds. Note that in the notation of the static model $y^*(i,i) = y^*_{i-1}$.

In summary, the structure of an optimal policy is as follows: Given d requests from customer class $i = 1, \ldots, i_{\max}$, a non-negative number of y^*_{i-1} seats (the so-called protection level of class $i-1$) is reserved for future demand of classes $i-1, \ldots, 1$. The protection levels are lower for higher value demand.

5.2.3 A Numerical Example

As an illustration, consider the following data taken from van Ryzin and McGill (2000): There are 4 fare classes with fare prices of $\varrho_1 = 1050 \geq \varrho_2 = 567 \geq \varrho_3 = 527 \geq \varrho_4 = 350$. The total capacity is $C = 100$. The demand is normally distributed (rounded to non-negative integer values with $d_{\max} = 500$). Table 5.2 shows the associated expectations $\mathbb{E}[D_i]$ and standard deviations $\sigma[D_i]$.

The protection levels of the optimal policy in the risk-neutral setting (obtained by solving (5.5)) read: $y^*_3 = 133$, $y^*_2 = 44$, and $y^*_1 = 17$. This means that e.g. no seats are sold to class 4 customers because $C < 133$. For classes 1 and 2, 44 seats are protected. So at most $100 - 44 = 56$ seats could be sold to class 3 customers. Accordingly, 17 seats are protected for late-arriving class 1 requests. The expected revenue is about 60038.

Table 5.2. Expectations and standard deviations of the number of requests.

fare class i	$\mathbb{E}[D_i]$	$\sigma[D_i]$
1	17.3	5.8
2	45.1	15.0
3	73.6	17.4
4	19.8	6.6

5.2.4 The EMSR Heuristics

Although the computation of optimal controls for the static (single-resource) capacity control model is not particularly difficult, exact optimization models are not widely used in practice. According to Talluri and van Ryzin (2004b, p. 45), most (single-resource) airline revenue management systems use heuristics to compute protection levels, "because they are simpler to code, quicker to run, and generate revenues that in many cases are close to optimal."

The two most popular heuristics for finding near-optimal protection levels $y_1, \ldots, y_{i_{\max}-1}$ are Belobaba's expected marginal seat revenue (EMSR) heuristics in versions a (EMSR-a) and b (EMSR-b). For EMSR-a, see Belobaba (1987a), (1987b), and (1989); for EMSR-b, see Belobaba and Weatherford (1996). Both heuristics approximate the $i_{\max}$ class static single-resource revenue management model given a risk-neutral, i.e. expected revenue maximizing, decision-maker by extending Littlewood's (1972) two-class solution to the $i_{\max}$ class model. They differ only in how they extend the solution method.

Littlewood's Two-Class Model

For only $i_{\max} = 2$ fare classes, the static capacity control problem is very similar to the newsvendor problem that dates back to Edgeworth (1888); see Petruzzi and Dada (1999) for a review. Pfeifer (1989) emphasizes the similarity between the newsvendor problem and the basic static model.

The observed class 2 demand d_2, which can be interpreted as d_2 requests for one of the C seats in the airplane, the probability distribution of class 1 demand, and the fares $\varrho_1 > \varrho_2$ are given. The question is how much class 2 demand to accept before seeing the realization of class 1 demand. Stated differently, it must be decided how many seats to protect from class 2 demand for future class 1 demand.

A simple marginal analysis motivates the optimal solution: Suppose there are c units of capacity remaining and there is an additional class 2 request. If the request is accepted, the airline collects a revenue of ϱ_2 for this unit of capacity. If it is not accepted, the unit can be sold at ϱ_1 if and only if class 1 demand is c or higher, i.e. if $D_1 \geq c$. Therefore, the expected marginal seat revenue from reserving the cth unit for class 1, $EMSR_1(c)$, is $\varrho_1 P(D_1 \geq c)$.

One should accept more class 2 demand as long as its price exceeds this marginal value.

If the expected marginal seat revenue of seat y and class i is defined as

$$EMSR_i(y) = P(D_i \geq y)\varrho_i \ ,$$

the optimal protection level, i.e. number of protected seats, y_1^* satisfies

$$EMSR_1(y_1^*) > \varrho_2 \text{ and } EMSR_1(y_1^* + 1) \leq \varrho_2 \ . \tag{5.7}$$

In the case of two booking classes, the expected marginal seat revenue of seat y and class 1 expresses the expected revenue gain by offering seat y to class 1. The optimal protection level y_1^* is the largest value of y for which the expected marginal seat revenue of class 1 is higher than the class 2 fare.

The EMSR-a Heuristic

In his thesis, Belobaba (1987a, pp. 126–131) proposes a straightforward extension of the two-class solution to the $i_{\max}$ class model, which is known as the EMSR-a heuristic. The protection level for the current class i with revenue ϱ_i is determined by summing up all the protection levels relative to each of the higher classes.

Assume one is interested in y_{i-1}, i.e. the number of seats to protect from class i or, stated differently, to reserve for future classes $i-1, \ldots, 1$. In order to determine y_{i-1}, pick a single class j among the future classes $i-1, \ldots, 1$ and compare i and j in isolation first. Considering only these two classes, (5.7) is used and y_{i-1}^j seats are reserved for class j, where

$$EMSR_j(y_{i-1}^j) > \varrho_i \text{ and } EMSR_j(y_{i-1}^j + 1) \leq \varrho_i \ . \tag{5.8}$$

Repeating this for each future class $j = i-1, \ldots, 1$ yields how much capacity to protect from class i for each class j in isolation. The idea of EMSR-a is to add up these individual protection levels to approximate the total protection level y_{i-1} as

$$y_{i-1}^a = \sum_{j=1}^{i-1} y_{i-1}^j \ . \tag{5.9}$$

According to Talluri and van Ryzin (2004b, p. 46), EMSR-a was believed to be optimal for a short time because of its intuitive appeal, but this was disproved by the published work on optimal controls. In general, EMSR-a can over- and underestimate optimal protection levels (see e.g. Brumelle and McGill (1993)). It performs extremely badly when there is a large number of classes with similar fare prices; for an example, see Talluri and van Ryzin (2004b, pp. 46–47) or Robinson (1995). Yet it requires less computational effort and is often only marginally worse than the optimal policy (see Robinson, 1995).

In the numerical example from Sect. 5.2.3, EMSR-a yields protection levels of $y_3^a = 106$, $y_2^a = 40$, and $y_1^a = 17$. Unsurprisingly, $y_1^a = y_1^*$, since the heuristic is exact given only two booking classes. The expected revenue generated by these protection levels (rounded to an integer value) is 60010, or 0.05% lower than optimal.

The EMSR-b Heuristic

Expected marginal seat revenue in version b is an alternative heuristic for the static $i_{\max}$ class model that reduces the problem for each period to the two-class model. In contrast to EMSR-a, the idea is to aggregate demand rather than protection levels.

EMSR-b aggregates demand from future classes and treats them as one artificial class in which demand is equal to the sum of the demands and revenue is equal to the weighted-average revenue of these classes. In so doing, it assumes that a passenger displaced by an additional booking would request a seat equal in revenue to this weighted average.

Let us again consider the calculation of the number of seats to protect from class i, y_{i-1}. EMSR-b considers only two classes, class i with associated revenue ϱ_i and the artificial class $(i-1)^\circ$. We define the demand of class $(i-1)^\circ$ as the aggregated future demand for classes $i-1, \ldots, 1$,

$$D_{(i-1)^\circ} = \sum_{j=1}^{i-1} D_j \, , \tag{5.10}$$

and the revenue $\varrho_{(i-1)^\circ}$ associated with class $(i-1)^\circ$ as the weighted average revenue from classes $1, \ldots, i-1$,

$$\varrho_{(i-1)^\circ} = \frac{\sum_{j=1}^{i-1} \varrho_j \mathbb{E}\left[D_j\right]}{\sum_{j=1}^{i-1} \mathbb{E}\left[D_j\right]} \, . \tag{5.11}$$

For these two classes, i and $(i-1)^\circ$, formula (5.7) can be used to determine $y_{(i-1)^\circ}$, the number of seats reserved for class $(i-1)^\circ$:

$$EMSR_{(i-1)^\circ}(y_{(i-1)^\circ}) > \varrho_i \text{ and } EMSR_{(i-1)^\circ}(y_{(i-1)^\circ} + 1) \le \varrho_i \, . \tag{5.12}$$

EMSR-b then sets the recommended protection level y_{i-1}^b equal to $y_{(i-1)^\circ}$.

Similar to EMSR-a, the EMSR-b heuristic has strong intuitive appeal. However, it assumes that the expected revenue of a future accepted request is equal to the expected revenue of all future requests. This is an approximation, because protection levels will also be set for higher booking classes, so the expected demand accepted in classes $j = i-1, \ldots, 1$ will not be equal to $\mathbb{E}[D_j]$ in general.

According to Talluri and van Ryzin (2004b, p. 48), EMSR-b is more frequently implemented in practice and generally performs better than EMSR-a, although simulation studies comparing the two are ambiguous. Talluri and van Ryzin (2004b) cite studies of Belobaba (1992) showing that EMSR-b consistently performs within 0.5 percent of the optimal revenue, whereas in certain cases EMSR-a deviated by nearly 1.5 percent from the optimal revenue. A study by Pölt (1999) showed a more mixed performance, with neither heuristic dominating the other.

In the numerical example from Sect. 5.2.3, EMSR-b yields protection levels of $y_3^b = 131$, $y_2^b = 51$, and $y_1^b = 17$. Unsurprisingly, $y_1^b = y_1^*$, since the heuristic is exact given only two booking classes. The expected revenue generated by these protection levels is about 59902, or 0.24% lower than optimal.

The results from solving the optimality equation, EMSR-a, and EMSR-b for the numerical example are listed in Table 5.3.

Table 5.3. Protection levels and expected revenue from solving the exact model, EMSR-a, and EMSR-b.

	y_3	y_2	y_1	$\mathbb{E}[R]$
exact	133	44	17	60038
EMSR-a	127	40	17	60010
EMSR-b	131	51	17	59902

Part III

Expected Utility Maximizing Capacity Control

The following chapters introduce the concept of risk-aversion to capacity control models. The static and the dynamic (capacity control) model are recapitulated from the perspective of an expected utility maximizing decision-maker. As indicated in Chap. 3, we restrict ourselves to the case of a decision-maker with either an additive time-separable utility function or an atemporal utility function.

Chapter 6 deals with the case of a decision-maker with an additive time-separable utility function. Given this preference structure, we show that nested protection levels are suitable for the dynamic model but not for the static model. In the dynamic model, we analyze the monotonicity of these protection levels in time and booking class as well as in the degree of risk-aversion. In the static model, optimal controls are monotone in the remaining capacity, yet they are not of protection level type. We also examine structures of an optimal policy with respect to the booking class or degree of risk-aversion. Our findings are illustrated by numerical examples.

After a critical discussion of the assumption of an additive time-separable utility function for capacity control problems, the structure of an optimal policy given an atemporal utility function is analyzed in Chap. 7. By means of examples, we show that many structural results known from the risk-neutral setting do not hold for a general shape of the utility function. Given constant absolute risk-aversion γ, i.e. an exponential utility function, however, all structural results of the expected revenue maximizing policy can be shown to hold for the expected utility maximizing policy in both the static and the dynamic problem. Under certain circumstances, the risk-neutral case can be shown to emerge as a special case for $\gamma \to 0$. The EMSR heuristics for the static model can be extended straightforwardly to account for this type of risk-aversion. In a small simulation study, we examine the impact of an expected atemporal exponential utility maximizing policy in a static model when the exact approach is applied or the heuristic counterparts are used. We compare our approach to the one proposed by Weatherford (2004). This chapter contains and extends the results of Barz and Waldmann (2007) and Barz (2006).

In Chap. 8, we examine how structures known from the risk-neutral setting also carry over to the case of a decision-maker maximizing expected atemporal exponential utility for more advanced, non-basic models, such as the capacity control model under a general discrete choice model of consumer behavior. We recapitulate this model as introduced by Talluri and van Ryzin (2004a) in the risk-neutral case. In a second step, we take the perspective of an expected utility maximizing decision-maker. By doing so, we concentrate on the case of an atemporal exponential utility function. This part reformulates and extends the ideas of Barz and Schön (2006).

In our analysis, we always assume the decision-maker's utility function to be known. For details on how to assess a decision-maker's utility function, see e.g. Keeney and Raiffa (1976, Sect. 4.9) or Eisenführ and Weber (2003, Sect. 9.4).

6

Capacity Control Maximizing Additive Time-Separable Utility

As mentioned before, the assumption of an additive time-separable utility function $\tilde{u}(\vec{\wr}) = \sum_{n=0}^{N} u_n(\wr^n)$, $u_n(0) = 0$ for all time periods $n = 0, \ldots, N$, is the one most frequently used in combination with Markov decision processes. Yet, as indicated in Chap. 3, it imposes a special structure of temporal and risk preferences.

In this section, we analyze the impact of additive time-separable preferences on the structure of an optimal control policy for the dynamic as well as the static capacity control problem. We assume the one-stage utility functions u_n to be increasing and concave to model a risk-averse decision-maker.

In both the dynamic and the static model, the terminal reward is chosen such that overbooking is not advisable in the absence of cancelations or no-shows. Every reasonable, non-overbooking policy has a terminal reward of zero. That is why we assume throughout that $u_0(\wr^0) = 0$ for $\wr^0 > 0$ and $u_0(\wr^0) = \wr^0 \bar{\varrho}^u / \bar{\varrho}$ else with $\bar{\varrho}^u > \max_{n=1,\ldots,N} u_n(\varrho_1)$. Of course, other shapes of u_0 can be imagined. But since the only purpose of the terminal reward is to ensure that at most C requests are accepted in total, all increasing functions u_0 with $u_0(\wr^0) = 0$ for $\wr^0 \geq 0$ that prevent overbooking result in the same add-optimal policy. So the only restriction imposed by this choice is that we do not account for scenarios in which negative capacity at flight departure, $n = 0$, is desirable.

6.1 The Dynamic Model

Applying the definition of $V^{\text{add}*}$ and the optimality equation (3.8) to the dynamic capacity control problem yields that the maximum expected (additive time-separable) utility starting with capacity c given a class i request N periods before departure,

$$V^{\text{add}*}(c,i) = \max_{\pi \in \mathfrak{F}^N} \mathbb{E}_\pi \left[\sum_{n=1}^{N} u_n\left(r_n(X_n, f_n(X_n))\right) + u_0\left(V_0(X_0)\right) \mid X_N = (c,i) \right]$$

for $(c,i) \in \mathfrak{X}$, is the unique solution $V^{\text{add}*} \equiv V_N^{\text{add}}$ to the optimality equation

$$V_n^{\text{add}}(c,i) = \max_{a \in \mathfrak{A}(c,i)} \left\{ u_n(a\varrho_i) + \sum_{i'=0}^{i_{\max}} \hat{p}_{n-1}(i') V_{n-1}^{\text{add}}(c-a,i') \right\} . \quad (6.1)$$

It can be obtained for $n = 1, \ldots, N$ iteratively, starting with the terminal reward $u_0(V_0(c,i))$ and u_0 as defined above. Every policy $\pi^{\text{add}*}$ formed by actions $a = f_n^{\text{add}*}(x)$, each maximizing the right-hand side of (6.1), is add-optimal, i.e. leads to $V^{\text{add}*}$.

Using $A_n v(c) = \sum_{i=0}^{i_{\max}} \hat{p}_n(i) v(c,i)$, the optimality equation (6.1) can also be stated as

$$A_n V_n^{\text{add}}(c) = \sum_{i=0}^{i_{\max}} \hat{p}_{n-1}(i) \max_{a \in \mathfrak{A}(c,i)} \left\{ u_n(a\varrho_i) + A_{n-1} V_{n-1}^{\text{add}}(c-a) \right\} ,$$

with initial value $A_0 V_0^{\text{add}}(c)$.

6.1.1 Structural Results

Since we are dealing with monetary outcomes, we assume $u_n(\wr)$ to be increasing in $\wr$ with $u_n(0) = 0$ for all $n = 0, \ldots, N$. The set of admissible actions $\mathfrak{A}(c,i)$ only contains the numbers 0 and 1 in this model. Thus, the utility functions simply rescale the fares ϱ_i to $u_n(\varrho_i)$. This rescaling of fares leaves many structural properties unchanged.

Substituting the fares ϱ_i by their utilities $u_n(\varrho_i)$, the following lemma follows directly from Lemma 5.1.

Lemma 6.1. *For $n = 1, \ldots, N$, $i = 0, \ldots, i_{\max}$ the value function $V_n^{add}(c,i)$ is increasing and concave in c.*

From this lemma and $\varrho_0 = 0$,

$$V_n(c,0) = A_{n-1} V_{n-1}(c) \geq A_{n-1} V_{n-1}(c-1) , \quad n = 1, \ldots, N, \; c \in \mathfrak{C}$$

follows. Thus, the set of feasible actions can be extended to $\mathfrak{A}$ for all $(c,i) \in \mathfrak{X}$ without loss of generality.

Similar to the expected revenue maximizing case, one can show that there is an add-optimal policy of the following structure.

Theorem 6.1. *Assume a decision-maker with additive time-separable utility function as defined above. There then exists an add-optimal policy $\pi^{add*} = (f_N^*, f_{N-1}^*, \ldots, f_1^*)$ for the dynamic capacity control problem such that*

$$f_n^*(c,i) = \begin{cases} 1 & c > y_{i-1}^{add*}(n) \\ 0 & c \leq y_{i-1}^{add*}(n) , \end{cases}$$

with time-dependent protection levels

$$y_{i-1}^{add*}(n) = \max\left\{c \in \{0,\dots,n-1\} : u_n\left(\varrho_i\right) < \Delta A_{n-1} V_{n-1}^{add}(c)\right\} .$$

For fixed n,

$$y_{i_{\max}-1}^{add*}(n) > y_{i_{\max}-2}^{add*}(n) > \cdots > y_0^{add*}(n) = 0 .$$

The proof is the same as in the dynamic model in Sect. 5.1.

For the risk-neutral setting, we were able to show that add-optimal protection levels are increasing in the number of periods until departure and never increase by more than one from period to period. This is certainly also true in the additive time-separable case if $u_n(\wr) = u_1(\wr)$ for all $n = 1,\dots,N$. Given different utility functions for different periods, however, the protection levels need not be increasing in n, and if they increase, it might well be by more than one. An example is given in the next subsection (Example 1).

Of course, a linear transformation of the utility function (with positive slope) leaves the add-optimal policy unchanged. Non-linear transformations of one-stage utility functions, however, can change the protection levels. If the decision-maker becomes more risk-averse within every period in the sense that all one-stage utility-functions are transformed by the same concave function g, protection levels do not increase.

Proposition 6.1. *Let $g : \mathbb{R} \to \mathbb{R}$ be a concave, increasing, and invertible function with $g(0) = 0$. The protection levels of a decision-maker with preferences represented by the utility function $u^1(\wr) = \sum_{n=0}^{N} u_n(\wr)$ are never lower than those of a decision-maker with preferences represented by $u^2(\wr) = \sum_{n=0}^{N} g(u_n(\wr))$.*

Proof. Let $V_n^{\text{add1}}(c,i)$ and $V_n^{\text{add2}}(c,i)$ denote the value functions of decision-makers with utility functions u^1 and u^2. From Theorems 6.1 and 5.1, we know that for both decision-makers an add-optimal policy can be described in terms of time-dependent protection levels

$$y_{i-1}^{\text{add1}*}(n) = \max\left\{c \in \mathbb{N}_0 : u_n\left(\varrho_i\right) < \Delta A_{n-1} V_{n-1}^{\text{add1}}(c)\right\} ,$$
$$y_{i-1}^{\text{add2}*}(n) = \max\left\{c \in \mathbb{N}_0 : g(u_n\left(\varrho_i\right)) < \Delta A_{n-1} V_{n-1}^{\text{add2}}(c)\right\} .$$

It must now be proven that $y_{i-1}^{\text{add1}*}(n) \geq y_{i-1}^{\text{add2}*}(n)$ for all $i = 1,\dots,i_{\max}$ and $n = 1,\dots,N$. As a first step, we establish the following two assertions:

(i) For all $i = 1,\dots,i_{\max}$, $n = 0,\dots,N-1$ and $c \geq 0$,

$$V_n^{\text{add2}}(c,i) - V_n^{\text{add2}}(c-1,i) \leq g\left(V_n^{\text{add1}}(c,i) - V_n^{\text{add1}}(c-1,i)\right) ;$$

(ii) For all $n = 0,\dots,N-1$ and $c \geq 0$,

$$\Delta A_n V_n^{\text{add2}}(c) \leq g\left(\Delta A_n V_n^{\text{add1}}(c)\right) .$$

We show (i) and (ii) by induction on n. For $n = 0$, (i) is a direct consequence of the definition of u_0, because g is concave and $g(0) = 0$. Given (i) is true for some n, assertion (ii) follows from

$$\begin{aligned}
\Delta A_n V_n^{\text{add2}}(c) &= \sum_{i=0}^{i_{\max}} \hat{p}_n(i)(V_n^{\text{add2}}(c,i) - V_n^{\text{add2}}(c-1,i)) \\
&\leq \sum_{i=0}^{i_{\max}} \hat{p}_n(i) g(V_n^{\text{add1}}(c,i) - V_n^{\text{add1}}(c-1,i)) \\
&\leq g\left(\sum_{i=0}^{i_{\max}} \hat{p}_n(i)(V_n^{\text{add1}}(c,i) - V_n^{\text{add1}}(c-1,i))\right) \\
&= g\left(\Delta A_n V_n^{\text{add1}}(c)\right) .
\end{aligned}$$

Now assume that (i) and (ii) hold for some $n \geq 0$. For $n+1$, we then obtain

$$\begin{aligned}
&V_{n+1}^{\text{add2}}(c,i) - V_{n+1}^{\text{add2}}(c-1,i) \\
&= \max\{g(u_{n+1}(\varrho_i)), \Delta A_n V_n^{\text{add2}}(c)\} \\
&\quad - \max\{g(u_{n+1}(\varrho_i)), \Delta A_n V_n^{\text{add2}}(c-1)\} + \Delta A_n V_n^{\text{add2}}(c-1) \\
&= \begin{cases} \Delta A_n V_n^{\text{add2}}(c) & , g(u_{n+1}(\varrho_i)) \leq \Delta A_n V_n^{\text{add2}}(c) \\ \Delta A_n V_n^{\text{add2}}(c-1) & , \Delta A_n V_n^{\text{add2}}(c-1) \leq g(u_{n+1}(\varrho_i)) \\ g(u_{n+1}(\varrho_i)) & , \text{otherwise} \end{cases} \\
&\leq \begin{cases} g\left(\Delta A_n V_n^{\text{add1}}(c)\right) & , u_{n+1}(\varrho_i) \leq \Delta A_n V_n^{\text{add1}}(c) \\ g\left(\Delta A_n V_n^{\text{add1}}(c-1)\right) & , \Delta A_n V_n^{\text{add1}}(c-1) \leq u_{n+1}(\varrho_i) \\ g(u_{n+1}(\varrho_i)) & , \text{otherwise} \end{cases} \\
&= g\left(V_{n+1}^{\text{add1}}(c,i) - V_{n+1}^{\text{add1}}(c-1,i)\right) ,
\end{aligned}$$

where we used the inequality $\Delta A_n V_n^{\text{add1}}(c-1) \geq \Delta A_n V_n^{\text{add1}}(c)$ from Lemma 6.1 and the fact that assertion (ii) holds for n.

An argument similar to the one used for $n = 0$ shows assertion (ii) given assertion (i). Hence, the induction is complete.

Applying (i) and (ii) to the protection levels given in Theorem 6.1 yields that for all $i = 1, \ldots, i_{\max}$ and $n = 1, \ldots, N$

$$\begin{aligned}
y_{i-1}^{\text{add2}*}(n) &= \max\left\{c \in \mathbb{N}_0 : g\left(u_n\left(\varrho_i\right)\right) < A_{n-1} V_{n-1}^{\text{add2}}(c)\right\} \\
&= \max\left\{c \in \mathbb{N}_0 : u_n\left(\varrho_i\right) < g^{-1}\left(A_{n-1} V_{n-1}^{\text{add2}}(c)\right)\right\} \\
&\leq \max\left\{c \in \mathbb{N}_0 : u_n\left(\varrho_i\right) < A_{n-1} V_{n-1}^{\text{add1}}(c)\right\} = y_{i-1}^{\text{add1}*}(n) .
\end{aligned}$$

Thus, $y_{i-1}^{\text{add2}*}(n) \leq y_{i-1}^{\text{add1}*}(n)$ for all $i = 1, \ldots, i_{\max}$ and $n = 0, \ldots, N-1$. □

6.1.2 Numerical Examples

We give two numerical examples to illustrate our results. The first one is to underline that protection levels need not be monotone in time given a

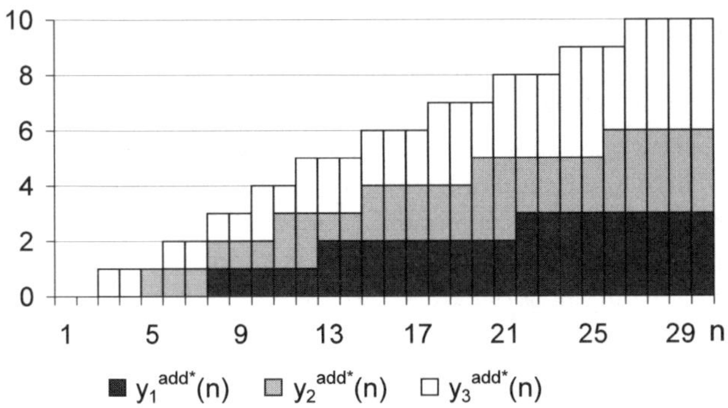

Fig. 6.1. Protection levels $y_i^{\text{add}*}(n)$ of an add-optimal policy given exponential one-stage utility functions with $\gamma = 0.002$.

decision-maker with a general, additive time-separable utility function. The second example reconsiders the numerical example given in the discussion of a risk-neutral, expected revenue maximizing decision-maker. The effect of introducing risk-aversion on the structure of an add-optimal policy is demonstrated.

Example 1: Non-Monotonicity in time

Consider a dynamic capacity control problem with $i_{\max} = 2$ fare classes $n = 4$ periods before departure. The fares are $\varrho_1 = 16$ and $\varrho_2 = 4$. The decision-maker maximizes expected utility and has an additive time-separable utility function with $u_1(\wr) = u_2(\wr) = u_4(\wr) = \wr^{1/2}$, $u_3(\wr) = \wr$. The number of seats that should be protected for class 1 customers over time can then be calculated as $y_1^{\text{add}*}(4) = 2$, $y_1^{\text{add}*}(3) = 0$, $y_1^{\text{add}*}(2) = 1$, $y_1^{\text{add}*}(1) = 0$. Clearly, these protection levels are non-monotone in time and vary by more than 1 from period to period (in contrast to the optimal protection levels in the case of a risk-neutral decision-maker).

Example 2: The Example Given by Lee and Hersh (1993) (Continued)

We continue the example from Sect. 5.1.3. Seats are sold in four booking classes with fares $\varrho_1 = 200$, $\varrho_2 = 150$, $\varrho_3 = 120$, and $\varrho_4 = 80$. There is a capacity of $C = 10$ seats $N = 30$ booking periods before departure. The request probabilities are listed in Table 5.1.

Assume now that the preferences of the decision-maker can be expressed by an additive time-separable utility function with exponential one-period utility functions $u_n(\wr) = 1 - \exp(-\gamma\wr)$, $n = 1, \ldots, N$. Note that 1 is added to the exponential utility formulation given in (3.4) to assure $u_n(0) = 0$ for all n.

Figures 6.1 and 6.2 show the time-dependent protection levels $y_{i-1}^{\text{add}*}(n)$ for $\gamma = 0.002$ and $\gamma = 0.01$. $y_1^{\text{add}*}(n)$ corresponds to the black columns, $y_2^{\text{add}*}(n)$ to the gray, and $y_3^{\text{add}*}(n)$ to the white. Recall that $n = 0$ represents flight departure. As shown in Theorem 6.1, the protection levels are increasing in i. In accordance with the result of Proposition 6.1, the protection levels decrease for higher single period risk-aversion. Therefore, the protection levels when $\gamma = 0.002$ are slightly lower than in the risk-neutral case considered in Sect. 5.1.3 and higher than those when $\gamma = 0.01$.

6.2 The Static Model

Assume that in static capacity control problem, the decision-maker is interested in finding

$$V^{\text{add}*}(c,d) = \max_{\pi \in \Pi} \mathbb{E}_\pi \left[\sum_{i=1}^{i_{\max}} u_i\left(r_i(X_i, f_i(X_i))\right) + u_0\left(V_0(X_0)\right) \mid X_{i_{\max}} = (c,d) \right],$$

the maximum expected (additive time-separable) utility starting with capacity c given d requests from class $i_{\max}$ over all policies $\Pi = \mathfrak{F}^{i_{\max}}$.

Using $A_i v(c) = \sum_{d=0}^{d_{\max}} \hat{p}_i(d) v(c,d)$, the optimality equations of the additive utility maximization problem can be stated as

$$\begin{aligned} V_i^{\text{add}}(c,d) &= \max_{a=0,\ldots,d} \{u_i(a\varrho_i) + \sum_{d'=0}^{d_{\max}} p_{i-1}(d') V_{i-1}^{\text{add}}(c-a,d) \\ &= \max_{a=0,\ldots,d} \{u_i(a\varrho_i) + A_{i-1} V_{i-1}^{\text{add}}(c-a)\} \end{aligned}$$

with terminal reward $V_0^{\text{add}}(c,d) = u_0(V_0(c,d)) = 0$ for $c \geq 0, d \in \mathfrak{D}$ and $V_0^{\text{add}}(c,d) = u_0(V_0(c,d)) = c\bar{\varrho}^u$ for $c < 0, d \in \mathfrak{D}$ where $\bar{\varrho}^u$ is sufficiently large to prevent overbooking, i.e. $\bar{\varrho}^u > \max_{i=1,\ldots,i_{\max}} u_i(\varrho_1)$.

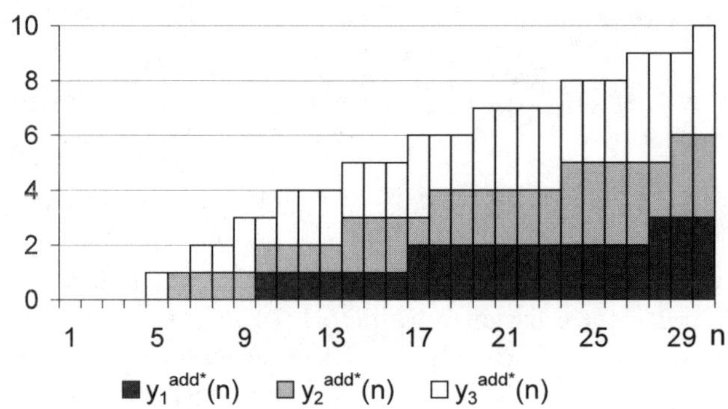

Fig. 6.2. Protection levels $y_i^{\text{add}*}(n)$ of an optimal policy given exponential one-stage utility functions with $\gamma = 0.01$.

Rearranging the optimality equation yields

$$\begin{aligned}V_i^{\mathrm{add}}(c,d) &= \max_{a=0,\ldots,d}\{u_i(a\varrho_i) + A_{i-1}V_{i-1}^{\mathrm{add}}(c-a) - A_{i-1}V_{i-1}^{\mathrm{add}}(c)\} \\ &\quad + A_{i-1}V_{i-1}^{\mathrm{add}}(c) \\ &= \max_{a=0,\ldots,d}\left\{\sum_{a'=1}^{a}\Big[u_i\big(a'\varrho_i\big) - u_i\big((a'-1)\varrho_i\big) \right. \\ &\quad \left. - \Delta A_{i-1}V_{i-1}^{\mathrm{add}}(c-a'+1)\Big]\right\} + A_{i-1}V_{i-1}^{\mathrm{add}}(c)\,. \qquad (6.2)\end{aligned}$$

The recursion given in (6.2) turns out to be advantageous for proving structural results.

Again, the equivalent form is

$$A_iV_i^{\mathrm{add}}(c) = \sum_{d'=0}^{d_{\max}} \hat{p}_i(d') \max_{a=0,\ldots,d}\{u_i(a\varrho_i) + A_{i-1}V_{i-1}^{\mathrm{add}}(c-a)\}\,.$$

6.2.1 Structural Results

The following lemma yields that all summands in (6.2) are increasing in c.

Lemma 6.2. *For fixed $i = 1,\ldots,i_{\max}$ and $d \in \mathbb{N}_0$, the value function $V_i^{add}(c,d)$ is increasing and concave in c.*

Proof. The proof is by induction on i. $V_0^{\mathrm{add}}(c,d)$ is increasing and concave in c. The values for $i > 0$ can be obtained by

$$V_i^{\mathrm{add}}(c,d) = \max_{a=0,1,\ldots,d}\{u_i(a\varrho_i) + A_{i-1}V_{i-1}^{\mathrm{add}}(c-a)\}\,.$$

Given $V_i^{\mathrm{add}}(c,d)$ is increasing and concave for some i, the linear combination $g(c) = A_{i-1}V_{i-1}^{\mathrm{add}}(c-a)$ is also increasing and concave. Thus, the maximum $V_{i+1}^{\mathrm{add}}(c,d)$ is increasing. Concavity follows from the induction hypothesis and the concavity of u_i using Lemma B.2 (given Appendix B), which is a straightforward extension of Lemma 1 in Stidham (1978). □

Again, the following lemma is obvious.

Lemma 6.3. *For $i \in \{1,\ldots,i_{\max}\}$ and $d \in \{0,\ldots,d_{\max}\}$, we have*

(i) $V_i^{add}(c,d) = u_i(\varrho_i d) + A_{i-1}V_{i-1}^{add}(c-1)$,
$= u_i(\varrho_i d) + \sum_{i'=1}^{i-1}\sum_{d'=0}^{d_{\max}} \hat{p}_{i'}(d')u_{i'}(\varrho_{i'}d')$, $c \geq id_{\max}$.
(ii) $V_i^{add}(c,d) = A_{n-i}V_{i-1}^{add}(c) = \bar{\varrho}^u c$, $c \leq 0$.

Proof. Both assertions follow by induction on i. The value function at stage $n = 1$ is

$$V_1^{\text{add}}(c,d) = \max_{a=0,\ldots,d} \left\{u_1(\varrho_1 a) + V_0^{\text{add}}(c-a,0)\right\} .$$

Using $V_0^{\text{add}}(c-a,0) = (c-a)\bar{\varrho}^u$ for $c \leq 0$, it follows that

$$u_1(\varrho_1 a) + V_0^{\text{add}}(c-a,0) = \bar{\varrho}^u c + u_1(\varrho_1 a) - \bar{\varrho}^u a .$$

This term is decreasing in a, because u_1 is concave. As $\bar{\varrho}^u > u_1(\varrho_1)$, action $a = 0$ maximizes the value function fulfilling (ii). For $c \geq d_{\max}$, every action results in a non-negative capacity at $i = 0$ with a terminal reward of 0. Consequently, the only term of interest is $u_1(\varrho_1 a)$, which is increasing in a by assumption. Hence, the maximum is attained for $a = d$ and (i) is shown.

Assume that (i) and (ii) are true for some $1 \leq i < i_{\max}$. For $c \leq 0$, $V_{i+1}^{\text{add}}(c,d)$ can be rearranged to read

$$V_{i+1}^{\text{add}}(c,d) = \max_{a=0,\ldots,d} \{\bar{\varrho}^u c + u_{i+1}(\varrho_{i+1} a) - \bar{\varrho}^u a .\} = \bar{\varrho}^u c ,$$

where the maximum is attained for $a = 0$, showing (ii). For $c \geq (i+1)d_{\max}$, every action causes a transition to a state $c - a \geq i d_{\max}$. $V_{i+1}^{\text{add}}(c,d)$ can be written as

$$V_{i+1}^{\text{add}}(c,d) = \max_{a=0,\ldots,d} \left\{ u_{i+1}(\varrho_{i+1} a) + \sum_{i'=1}^{i} \sum_{d'=0}^{d_{\max}} \hat{p}_{i'}(d') u_{i'}(\varrho_{i'} d') \right\} .$$

Again, $a = d$ maximizes the right-hand side, showing (i). □

We are now in a position to state the following property of an add-optimal policy:

Theorem 6.2. *Assume a decision-maker who maximizes expected additive time-separable utility with increasing and concave one-stage utility functions* u_i*. There then exists an add-optimal policy* $\pi^{add*} = (f_N^*, f_{N-1}^*, \ldots, f_1^*)$ *for the static capacity control problem such that*

$$f_i^{add*}(c,d) = \begin{cases} \min\{d, c - y_{i-1}^{add*}(c)\} & c > y_{i-1}^{add*}(c) \\ 0 & c \leq y_{i-1}^{add*}(c) , \end{cases}$$

with capacity dependent controls

$$y_{i-1}^{add*}(c) = \max\left\{ y \in \{0,\ldots,d_{\max}\} : u_i\left((c-y+1)\varrho_i\right) - u_i\left((c-y)\varrho_i\right) < \Delta A_{i-1} V_{i-1}^{add}(y) \right\}$$

and $y_0^{add*}(c) = 0$*. In addition, for all* i *and* c*,*

$$0 \leq y_{i-1}^{add*}(c+1) - y_{i-1}^{add*}(c) \leq 1 .$$

Proof. From Lemma 6.2, we can conclude that $\Delta A_{i-1} V_{i-1}^{\text{add}}(c-\alpha+1)$ is non-negative and decreasing in c, and that u is concave by assumption. Thus, the summands in (6.2) are decreasing in a'. It follows directly from (6.2) that one sells more and more seats, i.e. increases a, as long as there is demand and

$$u_i(a\varrho_i) - u_i((a-1)\varrho_i) - \Delta A_{i-1} V_{i-1}^{\text{add}}(c-a+1) > 0$$

holds. Combined with Lemma 6.3, this yields the controls $y_{i-1}^{\text{add}*}(c)$ stated above. $y_0^{\text{add}*}(c) = 0$ holds because $u_1(\varrho_1) > 0$, $A_0 V_0^{\text{add}}(y) = 0$ for all $y > 0$, and $A_0 V_0^{\text{add}}(0) = \bar{\varrho}^u > u_1(\varrho_1)$ by definition of $\bar{\varrho}^u$.

Another immediate consequence of Lemma 6.2 and equation (6.2) is that the add-optimal action is increasing in c. Consequently, $c+1-y_{i-1}^{\text{add}*}(c+1) \geq c - y_{i-1}^{\text{add}*}(c)$ or

$$y_{i-1}^{\text{add}*}(c+1) \leq y_{i-1}^{\text{add}*}(c) + 1 \ .$$

Since all utility functions u_i are assumed to be concave, it holds that

$$\begin{aligned} &u_i\left(((c+1)-y+1)\varrho_i\right) - u_i\left(((c+1))\varrho_i\right) \\ &\leq u_i\left((c-y+1)\varrho_i\right) - u_i\left((c-y)\varrho_i\right) \ . \end{aligned}$$

Applying this to the definition of the control $y_{i-1}^{\text{add}*}(c)$ yields

$$y_{i-1}^{\text{add}*}(c+1) \geq y_{i-1}^{\text{add}*}(c) \ ,$$

completing the proof. □

Note that given a decision-maker with additive time-separable utility function, there need not be an add-optimal policy that can be described in terms of protection levels (which are capacity-independent) for the static capacity control model. One can think of this as a consequence of the additive time-separable utility function that is composed of concave one-stage utility functions. The concave utility functions impose a preference for a smooth income stream over time and destroy the structure known from the risk-neutral setting.

In general, the add-optimal controls $y_i^{\text{add}*}(c)$ need not be increasing in the fare class i. Yet one can show the following sufficient condition.

Proposition 6.2. *If $u_i\left(a\varrho_i\right) - u_i\left((a-1)\varrho_i\right)$ is decreasing in i for all a, the controls $y_i^{add*}(c)$ are decreasing in i.*

The proposition is a direct consequence of the definition of $y_i^{\text{add}*}(c)$ and the following lemma.

Lemma 6.4. *$\Delta A_i V_i^{add}(c)$ is increasing in i for all $c = 0, \ldots, C$.*

Proof. From Theorem 6.2, we know that the add-optimal action in stage i given c seats and facing a demand d is

$$f_i^{\text{add}*}(c,d) = \min\{d, (c - y_{i-1}^{\text{add}*}(c))^+\} .$$

Thus, using

$$0 \le y_{i-1}^{\text{add}*}(c+1) - y_{i-1}^{\text{add}*}(c) \le 1$$

and introducing

$$\begin{aligned} h(c,i) = {} & u_i\left((c - y_{i-1}^{\text{add}*}(c))\varrho_i\right) - u_i\left((c-1-y_{i-1}^{\text{add}*}(c-1))\varrho_i\right) \\ & + A_{i-1}V_{i-1}^{\text{add}}\left(y_{i-1}^{\text{add}*}(c)\right) - A_{i-1}V_{i-1}^{\text{add}}\left(y_{i-1}^{\text{add}*}(c-1)\right) \end{aligned}$$

gives

$$\begin{aligned} \Delta A_i V_i^{\text{add}}(c) &= \sum_{d=0}^{d_{\max}} \hat{p}_i(d)\left[V_i^{\text{add}}(c,d) - V_i^{\text{add}}(c-1,d)\right] \\ &= \sum_{d=0}^{d_{\max}} \hat{p}_i(d)\Bigg[u_i\Big(\min\left\{d,(c - y_{i-1}^{\text{add}*}(c))^+\right\}\varrho_i\Big) \\ &\quad + A_{i-1}V_{i-1}^{\text{add}}\Big(c - \min\left\{d,(c - y_{i-1}^{\text{add}*}(c))^+\right\}\Big) \\ &\quad - u_i\Big(\min\left\{d,(c-1-y_{i-1}^{\text{add}*}(c-1))^+\right\}\varrho_i\Big) \\ &\quad - A_{i-1}V_{i-1}^{\text{add}}\Big(c-1-\min\left\{d,(c-1-y_{i-1}^{\text{add}*}(c-1))^+\right\}\Big)\Bigg] \\ &= \begin{cases} \Delta A_{i-1}V_{i-1}^{\text{add}}(c) & , c \le y_{i-1}^{\text{add}*}(c) \\ \sum_{d=0}^{c-1-y_{i-1}^{\text{add}*}(c)} \hat{p}_i(d)\Delta A_{i-1}V_{i-1}^{\text{add}}(c-d) & \\ \quad + \sum_{d=c-y_{i-1}^{\text{add}*}(c)}^{d_{\max}} \hat{p}_i(d) h(c,i) & , c > y_{i-1}^{\text{add}*}(c) . \end{cases} \end{aligned}$$

Our aim is to show that in both cases, $\Delta A_i V_i^{\text{add}}(c)$ is smaller than or equal to $\Delta A_{i-1}V_{i-1}^{\text{add}}(c)$. This is clearly true for the case $c \le y_{i-1}^{\text{add}*}(c)$. For $c > y_{i-1}^{\text{add}*}(c)$, we need to consider that $\Delta A_{i-1}V_{i-1}^{\text{add}}(c-d) \ge \Delta A_{i-1}V_{i-1}^{\text{add}}(c)$ (cf. Lemma 6.2) and that either $y_{i-1}^{\text{add}*}(c+1) = y_{i-1}^{\text{add}*}(c)$ or $y_{i-1}^{\text{add}*}(c+1) = y_{i-1}^{\text{add}*}(c)+1$ (cf. Theorem 6.2). If $y_{i-1}^{\text{add}*}(c+1) = y_{i-1}^{\text{add}*}(c)$, we obtain

$$\begin{aligned} h(c,i) &= u_i\left((c - y_{i-1}^{\text{add}*}(c))\varrho_i\right) - u_i\left((c-1-y_{i-1}^{\text{add}*}(c))\varrho_i\right) \\ &\ge \Delta A_{i-1}V_{i-1}^{\text{add}}\left(y_{i-1}^{\text{add}*}(c)+1\right) \\ &\ge \Delta A_{i-1}V_{i-1}^{\text{add}}(c) , \end{aligned}$$

where the first inequality follows from the definition of the protection levels and the second from Lemma 6.2 and $c > y_{i-1}^{\text{add}*}(c)$. Given $y_{i-1}^{\text{add}*}(c+1) = y_{i-1}^{\text{add}*}(c)+1$

$$\begin{aligned} h(c,i) &= \Delta A_{i-1}V_{i-1}^{\text{add}}\left(y_{i-1}^{\text{add}*}(c)\right) \\ &\ge \Delta A_{i-1}V_{i-1}^{\text{add}}(c) , \end{aligned}$$

where the inequality follows again from Lemma 6.2 and $c > y_{i-1}^{\text{add}*}(c)$.

Putting everything together yields $\Delta A_i V_i^{\text{add}}(c) \le \Delta A_{i-1}V_{i-1}^{\text{add}}(c)$. □

In contrast to the dynamic model, for the static model there need not exist an optimal policy that is monotone in the degree of risk-aversion. The reason for this is that given an expected utility maximizing decision-maker with additive time-separable preferences, the preference for a smooth income stream and the degree of risk-aversion increase at the same time. These two forces have an opposite influence on the preferred control.

As the add-optimal controls explicitly depend on the number of remaining seats, the EMSR-a and EMSR-b heuristics discussed in the risk-neutral setting cannot be transformed straightforwardly. Yet should an extension be required, an expected marginal additive time-separable seat utility (EMATU) of seat y and class j could be defined as

$$EMATU_j(y) = P(D_j \geq y)\left[u_j(y\varrho_j) - u_j((y-1)\varrho_j)\right] .$$

In (5.8) and (5.12), the values of $EMATU$ must be compared to $u_i((c-y+1)\varrho_i) - u_i((c-y)\varrho_i)$, a term that depends on c, in addition to the fare ϱ_i.

6.2.2 Numerical Examples

Two numerical examples are discussed to illustrate the structure of an add-optimal policy for the static capacity control problem. The first serves as a counterexample for monotonicity of the controls in the booking class. The data given in Sect. 5.2.3 is used in the second example. For a decision-maker with an exponential one-stage utility function, the results demonstrate the dependence of the add-optimal controls on the remaining capacity as well as the non-monotonicity in the coefficient of absolute risk-aversion.

Example 1: Non-Monotonicity in Booking Class

Given a total capacity of $C = 50$, tickets are sold in $i_{\max} = 4$ booking classes at fares $\varrho_1 = 1000$, $\varrho_2 = 101$, $\varrho_3 = 100$, and $\varrho_4 = 10$. The demands at each stage are assumed to be independent identically distributed with $\hat{p}_i(d) = 0.2$ for $d = 0, 1, 2$; $\hat{p}_i(d) = 0.1$ for $d = 3, 4$; $\hat{p}_i(d) = 0.05$ for $d = 5$; and $\hat{p}_i(d) = 0.01$ for $d = 6, \ldots, 20$. Given a risk-neutral decision-maker, the optimal protection levels are $y_1^* = y_2^* = 0$ and $y_3^* = 18$. In line with the results of Theorem 5.2, these controls are independent of c and increasing in the booking class i. For an expected utility maximizing decision-maker with exponential one-stage utility function $u_n(\wr) = 1 - \exp(-\gamma\wr)$ for all $n = 1, \ldots, N$ and $\gamma = 0.05$, controls of $y_1^{\text{add}*}(c) = 0$ for all c are still preferred. The number of seats that should be protected for classes 2 and 3, however, depends on the number of seats available at the corresponding stage.

Figure 6.3 shows a plot of the add-optimal controls $y_2^{\text{add}*}(c)$ and $y_3^{\text{add}*}(c)$. The controls of booking class 2 are plotted in gray, the controls of class 3 are white, and dotted columns indicate that both are equal. Although it is obvious that the add-optimal controls depend on c, it can be seen that they

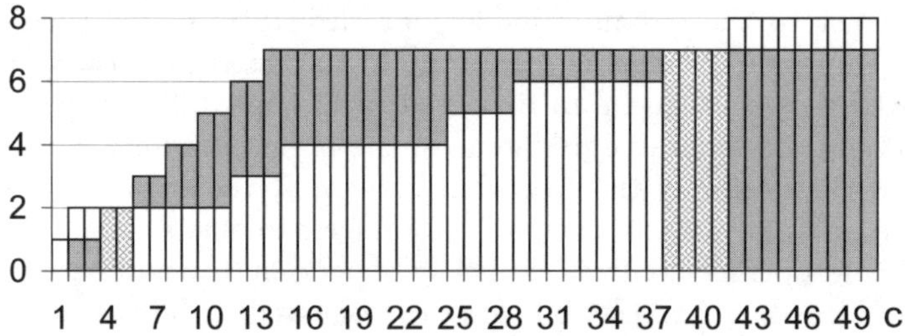

Fig. 6.3. Capacity dependent controls for classes 2 and 3 of an add-optimal policy given exponential one-stage utility functions with $\gamma = 0.05$.

never increase by more than 1. In addition, they need not be monotone in the booking class. In this example, $y_2^{\text{add}*}(c)$ is larger than $y_3^{\text{add}*}(c)$ for small values of c. They are equal for values of $38 \leq c \leq 41$. For higher values of c, the control $y_3^{\text{add}*}(c)$ is larger than $y_2^{\text{add}*}(c)$.

Example 2: The Example Given by van Ryzin and McGill (2000) (Continued)

We continue the example of Sect. 5.2.3 and concentrate on the case of $c = 85$ remaining seats. Consider a decision-maker with an additive time-separable utility function with identical exponential utility functions for every stage $n = 1, \ldots, N$. Figure 6.4 shows the add-optimal controls $y_1^{\text{add}*}(85)$, $y_2^{\text{add}*}(85)$, and $y_3^{\text{add}*}(85)$ for different values of the parameter $\gamma = 0.0001$, 0.0002, 0.0005, and 0.001. The black columns represent $y_1^{\text{add}*}(85)$, the gray ones $y_2^{\text{add}*}(85)$ and the white ones $y_3^{\text{add}*}(85)$. For increasing risk-aversion, the values of the add-optimal controls change in different directions. The controls $y_1^{\text{add}*}(85)$ increase, while the values of $y_2^{\text{add}*}(85)$ and $y_3^{\text{add}*}(85)$ decrease.

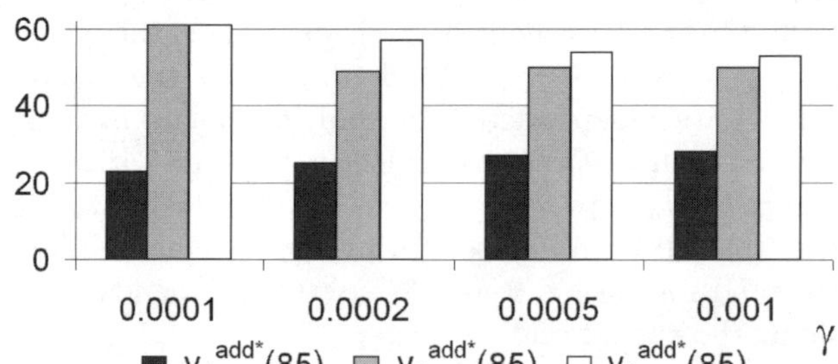

Fig. 6.4. Controls of an add-optimal policy for classes 1, 2 and 3 given 85 remaining seats and one-stage exponential utility functions with $\gamma = 0.0001, 0.0002, 0.0005$ and 0.001.

7

Capacity Control Maximizing Atemporal Utility

In the examples given in Sect. 6.2.2, the add-optimal policies had some rather counter-intuitive properties. Indeed, one might question the assumptions underlying the maximization of expected additive time-separable utility when incorporating risk-aversion into an MDP formulation.

In the dynamic capacity control model, the booking horizon is divided into time periods such that the probability of more than one arrival within one period is negligible. Consequently, the time periods considered are rather small. Given e.g. an airline setting, the time unit might even be hours or minutes. A desire for a certain shape of the income stream over these small time periods might be difficult to justify. Yet a risk-averse decision-maker maximizing expected additive time-separable utility has preferences concerning this shape.

At first sight, this argument is not valid for the static capacity control model, since the time periods considered are longer. Yet in most applications, the periods might be of very different length and are still so short that such temporal preferences seem unlikely. Indeed, very often some of the booking classes might even be open at the same time, and the protection levels determined by the static model are used simultaneously as a worst-case heuristic. (It is called worst-case because the order of arrival is the least preferred.) In this case, additive time-separable utility functions are inappropriate.

In the risk-neutral model, future revenues are usually not even discounted, so there is no hint at any type of temporal preference. This is justified by the fact that in typical applications, the booking horizon is relatively small, so the timing of the revenue gain is unimportant.

In this spirit, it seems more natural to assume a decision-maker with a utility function that evaluates the total revenue gained independent of the timing within the booking horizon. In this chapter, we analyze the dynamic as well as the static capacity control problem, with the aim of maximizing the expected utility of a decision-maker with an atemporal utility function.

7.1 The Dynamic Model

In the dynamic model, the maximum expected utility of total revenue observing a demand of class i and given c remaining seats as well as a current wealth of w is

$$V^{\text{atmp}*}(c,i,w) = \max_{\pi \in \Pi} \mathbb{E}_\pi \left[u(R_\pi) \mid (X_N, W_N) = (c,i,w) \right] ,$$

where u is the increasing utility function, and $W_n \in \mathfrak{W}$ is the additional wealth state as defined in Sect. 3.2.2. The total reward (added on the current wealth) is $R_\pi = W_N + \sum_{n=1}^{N} r_n(X_n, f_n(X_n, W_n)) + V_0(X_N)$. Note that the decision rule $f_n(X_n, W_n)$ might depend on the current wealth W_n as well as on the system state X_n.

According to the results stated in Sect. 3.2.2, $V^{\text{atmp}*}$ is the unique solution to the optimality equation

$$\begin{aligned} V_n^{\text{atmp}}(c,i,w) &= \max_{a \in \mathfrak{A}(c,i)} \left\{ \sum_{i'=0}^{i_{\max}} \hat{p}_{n-1}(i') V_{n-1}^{\text{atmp}}(c-a, i', w + a\varrho_i) \right\} \\ &= \max_{a \in \mathfrak{A}(c,i)} \left\{ \tilde{A}_{n-1} V_{n-1}^{\text{atmp}}(c-a, w + a\varrho_i) \right\} \end{aligned} \tag{7.1}$$

with

$$V_0^{\text{atmp}}(c,i,w) = u(V_0(c,i) + w)$$

and $\tilde{A}_n V_n^{\text{atmp}}(c,w) = \sum_{i=0}^{i_{\max}} \hat{p}_n(i) V_n^{\text{atmp}}(c,i,w)$. Every policy $\pi^{\text{atmp}*}$ formed by actions $a = f_n^{\text{atmp}*}(c,i,w)$, each maximizing the right-hand side of the optimality equation, is atmp-optimal, i.e. leads to $V^{\text{atmp}*}$.

In an implementation, this should be again be reformulated to

$$\tilde{A}_n V_n^{\text{atmp}}(c,w) = \sum_{i=0}^{i_{\max}} \hat{p}_n(i) \max_{\mathfrak{A}(c,i)} \left\{ \tilde{A}_{n-1} V_{n-1}^{\text{atmp}}(c-a, w + a\varrho_i) \right\} ,$$

starting from $\tilde{A}_0 V_0^{\text{atmp}}(c,w)$, in order to reduce the computational complexity.

7.1.1 Structural Results

The following lemma is straightforward to show.

Lemma 7.1. *$V_n^{atmp}(c,i,w)$ is increasing in c for all $i \in \mathfrak{I}$, $w \in \mathfrak{W}$, and $n = 0,\ldots,N$.*

Proof. The proof is by induction on n. First, the statement is true for $n = 0$ because u and $V_0(\cdot, i)$ are increasing. Assume it is true for some n. That would mean that $\tilde{A}_n V_n^{\text{atmp}}(c-1, w+\varrho_i)$ and $\tilde{A}_n V_n^{\text{atmp}}(c,w)$ are increasing in c. Therefore, the maximum, $V_{n+1}^{\text{atmp}}(c,i,w)$, is also increasing in c and the induction is complete. □

With $\varrho_0 = 0$ it follows that

$$V_n^{\text{atmp}}(c, 0, w) = \tilde{A}_{n-1} V_{n-1}^{\text{atmp}}(c, w) \geq \tilde{A}_{n-1} V_{n-1}^{\text{atmp}}(c-1, w) .$$

Thus, the set of feasible actions can be augmented without loss to $\mathfrak{A} = \{0, 1\}$ for all $(c, i) \in \mathfrak{X}$.

In order to show that there is an atmp-optimal policy that is monotone in the booking class i, we need the following lemma.

Lemma 7.2. *$V_n^{atmp}(c, i, w)$ is increasing in w for all $i \in \mathfrak{I}, c \in \mathfrak{C}$, and $n = 0, \ldots, N$.*

Proof. Again, the proof is by induction on n. The statement is true for $n = 0$ because u is increasing in w. Given it is true for some n, both $\tilde{A}_n V_n^{\text{atmp}}(c - 1, w + \varrho_i)$ and $\tilde{A}_n V_n^{\text{atmp}}(c, w)$, and thus their maximum, are increasing in w. □

As a consequence of $\tilde{A}_n V_n^{\text{atmp}}(c, w + \varrho_i) \geq \tilde{A}_n V_n^{\text{atmp}}(c, w)$, the inequality $\tilde{A}_n V_n^{\text{atmp}}(c, w + \varrho_{i'}) \geq \tilde{A}_n V_n^{\text{atmp}}(c, w)$ holds for all $1 \leq i' < i$. Stated differently, if a class i customer is accepted n periods before departure, then given a remaining capacity of c seats and a wealth of w, all higher-fare customers (classes $i' = 1, \ldots, i$) are also accepted.

The third lemma is obvious.

Lemma 7.3. *$\tilde{A}_n V_n^{atmp}(c, w)$ is increasing in n for all $c \in \mathfrak{C}$, and $w \in \mathfrak{W}$.*

Proof. The assertion follows directly from the optimality equation due to

$$\begin{aligned} \tilde{A}_n V_n^{\text{atmp}}(c, w) &= \sum_{i=0}^{i_{\max}} \hat{p}_n(i) \max_{a \in \mathfrak{A}(c,i)} \left\{ \tilde{A}_{n-1} V_{n-1}^{\text{atmp}}(c - a, w + a\varrho_i) \right\} \\ &\leq \tilde{A}_{n-1} V_{n-1}^{\text{atmp}}(c, w) . \quad \square \end{aligned}$$

7.1.2 Numerical Examples

In the special case of $C = 1$, the dynamic capacity control problem is an optimal stopping problem as discussed in Müller (2000) and Hall et al. (1979). Yet in general the total capacity is higher. We consider two examples in which $C > 1$. The first example is a toy example that shows that the atmp-optimal policy is in general neither monotone in c nor in w given an arbitrary increasing utility function u. The second is a continuation of the running example with the data given by Lee and Hersh (1993).

Example 1: Non-Monotonicity in c and w

Consider a decision-maker with a capacity of $C = 5$ seats several periods before departure. Customers request seats in $i_{\max} = 3$ booking classes with fares of $\varrho_1 = 5$, $\varrho_2 = 2$, and $\varrho_3 = 1$. The probability for a class 1 request n periods before departure is $\hat{p}_n(1) = 0.15$, for a class 2 request the probability is $\hat{p}_n(2) = 0.3$, and for a class 3 request it is $\hat{p}_n(3) = 0.3$. Hence, the probability of no customer request in period n is $\hat{p}_n(0) = 0.25$. Now assume that the decision-maker has a utility function with an aspiration level of $\beta = 6$.

Let us concentrate on atmp-optimal actions $n = 5$ periods before departure. Given a current wealth of $w = 0$, it is atmp-optimal to accept every incoming request with only one exception. If five periods before departure the remaining capacity is three, class 3 requests should be rejected. In particular, given a class 3 request five periods before departure, the atmp-optimal action is to sell one seat given $c = 2$, to sell zero seats given $c = 3$, and again to sell one given $c = 4$.

This can be explained as follows: If there are only two seats left, the only way of earning a total revenue of 6 or more is to sell one seat to a class 1 customer and one seat to any other class. An incoming class 3 request should therefore be accepted. If there are three seats left, there are more ways of earning at least 6 units of revenue within the next five periods. The question of whether to accept a class 3 request or not ends in a comparison of a "there will be three class 2 requests and one class 3 or no request within the next four time periods" scenario with "there will be one class 1 request and no other"; otherwise, the decision on this request has no impact on the probability of reaching the aspiration level. Since the first scenario is the more probable one, it is better to deny a class 3 request at $n = 5$. With four seats remaining, the comparison is between a "two class 2 and two class 3 requests" and a "two class 2, one class 3, and any other request", which turns out to favor accepting a class 3 customer.

Hence, there is no atmp-optimal policy that can be described by time-dependent protection levels; the atmp-optimal action is not even monotone in the remaining capacity.

If we consider the same decision-maker with a current wealth of $w = 1$, the combinatorial effects mentioned above cause the decision-maker to accept every request if the remaining capacity is two or more. Given one remaining seat, only the revenue from a class 1 customer will enable the decision-maker to reach his target. Consequently, an atmp-optimal policy is to reject requests from classes 2 and 3 for $c = 1$ and to accept them for $c > 1$.

Comparing this to the policy in which $w = 0$ illustrates that an atmp-optimal policy does not generally need to be monotone in w, because some atmp-optimal actions increase and others decrease.

Example 2: The Example Given by Lee and Hersh (1993) (Continued)

Let us again consider the example of Lee and Hersh (1993) as stated in Sect. 5.1.3. This time, however, we assume that the decision-maker maximizes expected utility given an atemporal utility function. If the utility function has an

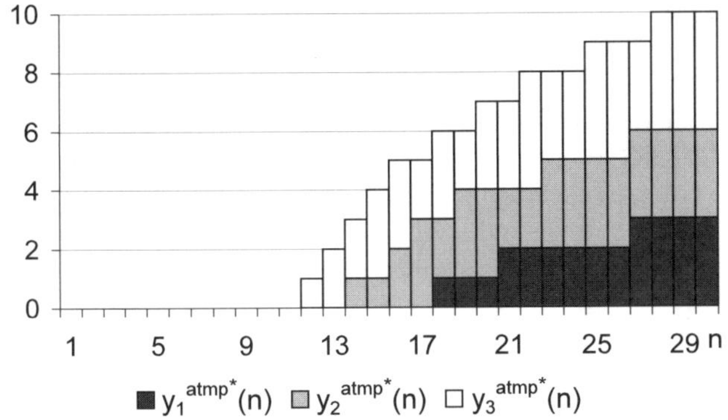

Fig. 7.1. Protection levels of an atmp-optimal policy given a decision-maker with logarithmic utility function and current wealth $w = 0$.

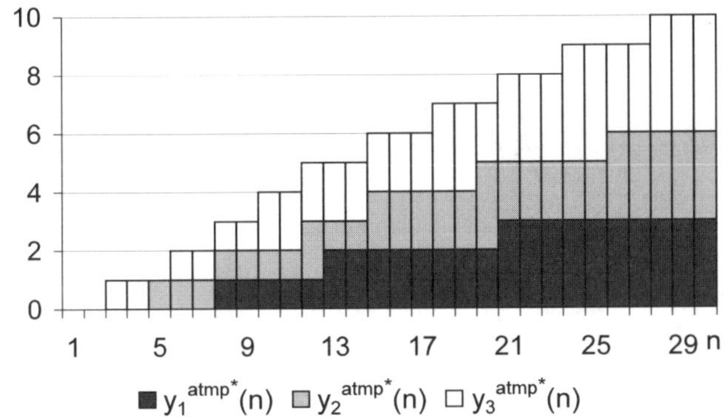

Fig. 7.2. Protection levels of an atmp-optimal policy given a decision-maker with logarithmic utility function and current wealth $w = 500$.

aspiration level of e.g. 1000, the decision problem turns into a combinatorial problem as in the first example.

Given the more well-behaved logarithmic utility function, however, there is an atmp-optimal policy for this example that can be described in terms of time-dependent protection levels. Atmp-optimal policies given a wealth of 0 and 500 are depicted in Figs. 7.1 and 7.2. $y_1^{\mathrm{atmp}*}(n)$, $y_2^{\mathrm{atmp}*}(n)$, and $y_3^{\mathrm{atmp}*}(n)$ are indicated in black, gray, and white, respectively. Monotonicity in time and capacity holds in addition to the proven monotonicity in fare class. It is also not surprising that these protection levels are increasing in w. Higher protection levels increase the risk of having empty seats and a relatively small total revenue at departure. As the degree of risk-aversion decreases in the wealth level, higher protection levels are tolerated in return for higher expected revenue for increasing w.

7.1.3 Structural Results in the Case of an Exponential Utility Function

The exponential utility function is the most widely used non-linear utility function and is said to approximate many types of risk-preferences satisfactorily (see Sect. 3.1.2). In addition, for an exponential utility function, the state space of the MDP can be reduced to $\mathfrak{X}$, which simplifies the analysis. This is why we concentrate on the special case of an exponential utility function in the following. Again, we write $A_n v(c)$ in place of $\sum_{i=0}^{i_{\max}} p_n(i) v(c,i)$ for an arbitrary real-valued function v on $\mathfrak{X}$.

Let $V^{\gamma *}(c,i)$, $(c,i) \in \mathfrak{X}$, denote the maximum expected atemporal exponential utility, i.e.

$$V^{\gamma *}(c,i) = \max_{\pi \in \Pi} \mathbb{E}_{\pi} \left[-\exp\left(-\gamma R_{\pi}\right) \mid X_N = (c,i) \right] . \tag{7.2}$$

The objective of finding a policy $\pi^{\gamma *} = (f_N^{\gamma *}, f_{N-1}^{\gamma *}, \ldots, f_1^{\gamma *})$, called γ-optimal, reduces to the solution of a risk-sensitive MDP. It follows that $V^{\gamma *} \equiv V_N^{\gamma}$ is the unique solution of

$$\begin{aligned} V_n^{\gamma}(c,i) &= \max_{a \in \mathfrak{A}(c,i)} \left\{ \exp(-\gamma a \varrho_i) \sum_{i'=0}^{i_{\max}} \hat{p}_{n-1}(i') V_{n-1}^{\gamma}(c-a, i') \right\} \\ &= \max_{a \in \mathfrak{A}(c,i)} \left\{ \exp(-\gamma a \varrho_i) A_{n-1} V_{n-1}^{\gamma}(c-a) \right\} \end{aligned} \tag{7.3}$$

for all $(c,i) \in \mathfrak{X}$, which can be obtained for $n = 1, \ldots, N$ by backward induction, starting with $V_0^{\gamma}(c,i) = -\exp(-\gamma V_0(c,i))$ for $(c,i) \in \mathfrak{X}$. Moreover, every policy $\pi^{\gamma *}$ formed by actions $a^{\gamma *} = f_n^{\gamma *}(c,i)$, each maximizing the right-hand side of (7.3), is γ-optimal, i.e. leads to $V^{\gamma *}$.

To prove structural results, it is more convenient to work with $G_n^{\gamma} := -V_n^{\gamma}$, which is the unique solution of

$$G_n^{\gamma}(c,i) = \min_{a \in \{0,1\}} \left\{ \exp(-\gamma a \varrho_i) A_{n-1} G_{n-1}^{\gamma}(c-a) \right\} \tag{7.4}$$

$$= A_{n-1} G_{n-1}^{\gamma}(c-1) \cdot \min \left\{ \exp(-\gamma \varrho_i), \; \frac{A_{n-1} G_{n-1}^{\gamma}(c)}{A_{n-1} G_{n-1}^{\gamma}(c-1)} \right\} \tag{7.5}$$

$$= A_{n-1} G_{n-1}^{\gamma}(c) \cdot \min \left\{ 1, \; \exp(-\gamma \varrho_i) \frac{A_{n-1} G_{n-1}^{\gamma}(c-1)}{A_{n-1} G_{n-1}^{\gamma}(c)} \right\} \tag{7.6}$$

with initial value $G_0^{\gamma} = -V_0^{\gamma}$. Note that (7.4) immediately follows from (7.3) by multiplication with (-1). The γ-optimality of a policy is preserved.

It is plausible that a γ-optimal policy accepts an arbitrary request if there is a remaining capacity c and merely $n \leq c$ periods remain. Furthermore, an arbitrary request is rejected in the case where $c \leq 0$. This is the result of the following lemma.

Lemma 7.4. *For $\gamma > 0$, $n \in \{1, \dots, N\}$, and $i \in \{1, \dots, i_{max}\}$, we have*

(i) $G_n^\gamma(c,i) = \exp(-\gamma\varrho_i) A_{n-1} G_{n-1}^\gamma(c-1)$
$= \exp(-\gamma\varrho_i) \prod_{n'=1}^{n-1} \sum_{i'=0}^{i_{max}} \hat{p}_{n'}(i') \exp(-\gamma\varrho_{i'}), \quad c \geq n.$

(ii) $G_n^\gamma(c,i) = A_{n-1} G_{n-1}^\gamma(c) = \exp(-\gamma\bar{\varrho}c), \quad c \leq 0.$

Proof. Using the inequalities $\exp(-\gamma\varrho_i) \leq 1 \leq \exp(\gamma(\bar{\varrho} - \varrho_i))$, (i) and (ii) follow by induction on n.

Both are certainly true for $n = 1$, since

$$G_1^\gamma(c,i) = \min\left\{\exp(-\gamma\varrho_i)\exp(-\gamma V_0(c-1,0)), \exp(-\gamma V_0(c,0))\right\} .$$

For $c \geq 1$, the terminal reward is $V_0(c,0) = 0$. Thus, (i) directly follows from $\exp(-\gamma\varrho_i) < 1$. For $c \leq 0$, the terminal reward is $V_0(c,0) = c\bar{\varrho}$, which results in

$$G_1^\gamma(c,i) = \min\left\{\exp(-\gamma c\bar{\varrho})\exp(\gamma(\bar{\varrho} - \varrho_i)), \exp(-\gamma c\bar{\varrho})\right\} = \exp(-\gamma c\bar{\varrho})$$

because $\bar{\varrho} > \varrho_1$. This ensures (ii) for $n = 1$.

Now assume both statements are true for some n. For $c \geq n+1$, it is straightforward to show that

$$\begin{aligned} G_{n+1}^\gamma(c,i) &= \min\left\{\exp(-\gamma\varrho_i) \prod_{n'=1}^{n} \sum_{i'=0}^{i_{\max}} \hat{p}_{n'}(i') \exp(-\gamma\varrho_{i'}),\right. \\ &\qquad \left. \prod_{n'=1}^{n} \sum_{i'=0}^{i_{\max}} \hat{p}_{n'}(i') \exp(-\gamma\varrho_{i'})\right\} \\ &= \exp(-\gamma\varrho_i) \prod_{n'=1}^{n} \sum_{i'=0}^{i_{\max}} \hat{p}_{n'}(i') \exp(-\gamma\varrho_{i'}) . \end{aligned}$$

The optimality equation for $c \leq 0$ at stage $n+1$ reduces to

$$\begin{aligned} G_{n+1}^\gamma(c,i) &= \min\left\{\exp(-\gamma(\varrho_i - \bar{\varrho}))\exp(-\gamma\bar{\varrho}c), \exp(-\gamma\bar{\varrho}c)\right\} \\ &= \exp(-\gamma\bar{\varrho}c) \end{aligned}$$

using the same argument as at stage $n = 1$. □

In the risk-neutral case, the value function is increasing and concave. The analogous result when maximizing expected atemporal exponential utility is the following.

Lemma 7.5. *For $\gamma > 0$, $n \in \{1, \dots, N\}$, and $i \in \{0, \dots, i_{max}\}$ it holds that*

(i) $A_{n-1} G_{n-1}^\gamma(c)$ *is log-convex and decreasing in* c.
(ii) $G_n^\gamma(c,i)$ *is log-convex and decreasing in* c.

Proof. The assertion follows by induction on n. Fix $\gamma > 0$. For all $1 \leq n \leq N$ set

$$g_{n-1}(c) = A_{n-1} G^{\gamma}_{n-1}(c), \quad c \in \mathbb{Z} .$$

Let $n = 1$. Since $-\gamma V_0(\cdot, j)$ is convex, and using the closure of log-convex functions with respect to convex combinations, we have log-convexity of g_0,

$$g_0(c) = \sum_{i'=0}^{i_{\max}} \hat{p}_0(i') \exp(-\gamma V_0(c, i')) .$$

As $V_0(\cdot, j)$ is increasing by assumption, g_0 is decreasing.

Next, we use g_0 to rewrite the optimality equation (7.4) as

$$\ln G^{\gamma}_1(c, i) = \min_{a \in \{0,1\}} \{-a\gamma \varrho_i + \ln g_0(c - a)\} . \tag{7.7}$$

Finally, by applying Lemma 1 in Stidham (1978) (see Appendix B) to (7.7), we obtain convexity of $\ln G^{\gamma}_1(\cdot, i)$, $i = 0, \ldots, i_{\max}$. Hence, $G^{\gamma}_1(\cdot, i)$ is not only decreasing (as $\ln G^{\gamma}_1(\cdot, i)$ is the minimum of decreasing functions) but also log-convex.

Now suppose $G^{\gamma}_n(\cdot, i)$ is log-convex for some $1 \leq n < N$. Then

$$g_n(c) = \sum_{i'=0}^{i_{\max}} \hat{p}_n(i') G^{\gamma}_n(c, i')$$

is decreasing and log-convex as a convex combination of decreasing and log-convex functions. By applying Stidham's lemma (1978) to

$$\ln G^{\gamma}_{n+1}(c, i) = \min_{a \in \{0,1\}} \{-a\gamma \varrho_i + \ln g_n(c - a)\} ,$$

we finally obtain the desired log-convexity of $G^{\gamma}_{n+1}(\cdot, i)$, $i = 0, \ldots, i_{\max}$. Monotonicity of $G^{\gamma}_{n+1}(\cdot, i)$ is a direct consequence of the monotonicity of g_n. □

From these lemmas, we immediately obtain the following theorem.

Theorem 7.1. *For the dynamic problem there exists a* γ*-optimal policy* $\pi^{\gamma *} = (f^{\gamma *}_N, f^{\gamma *}_{N-1}, \ldots, f^{\gamma *}_1)$ *such that*

$$f^{\gamma *}_n(c, i) = \begin{cases} 1 & c > y^{\gamma *}_{i-1}(n) \\ 0 & c \leq y^{\gamma *}_{i-1}(n) , \end{cases}$$

with time-dependent protection levels

$$y^{\gamma *}_{i-1}(n) = \max\left\{c \in \{0, \ldots, n-1\} \ : \ \exp(-\gamma \varrho_i) > \Gamma A_{n-1} G^{\gamma}_{n-1}(c)\right\} .$$

Proof. By Lemma 7.5 (i), $A_{n-1}G^{\gamma}_{n-1}(c)$ is log-convex in c. Therefore, $A_{n-1}G^{\gamma}_{n-1}(c)$ is a positive function, and

$$\Gamma A_{n-1}G^{\gamma}_{n-1}(c) = \frac{A_{n-1}G^{\gamma}_{n-1}(c)}{A_{n-1}G^{\gamma}_{n-1}(c-1)}$$

is increasing in c. Together with Lemma 7.4, we may then define constants $y^{\gamma*}_{i-1}(n)$ as defined above for $i = 1, \ldots, i_{\max}$. These constants determine γ-optimal controls according to (7.4). □

From Lemma 7.2, we know that the protection levels of an atmp-optimal policy are increasing in the booking class for an increasing utility function u. Hence, this is also true in case of an exponential utility function. Additionally, we are now in a position to show monotonicity in the remaining number of time periods.

Proposition 7.1. *The protection levels $y^{\gamma*}_{i-1}(n)$ of a γ-optimal policy satisfy*

(i) $y^{\gamma*}_{i-1}(n)$ is increasing in $i = 1, \ldots, i_{max}$ with $y^{\gamma*}_{0}(n) = 0$ for all $n = 1, \ldots, N$ and $\gamma > 0$,
(ii) $y^{\gamma*}_{i-1}(n-1) \leq y^{\gamma*}_{i-1}(n) \leq y^{\gamma*}_{i-1}(n-1) + 1$ for all $n = 2, \ldots, N$, $i \in \Im$ and $\gamma > 0$.

Proof. We prove the assertions in reverse order. To prove (ii), use (7.5) and Lemma 7.5 (i) in order to obtain

$$\begin{aligned} \Gamma A_n V^{\gamma}_n(c) =& \Gamma A_{n-1}V^{\gamma}_{n-1}(c-1) \\ & \cdot \frac{\sum_{i=0}^{i_{\max}} \hat{p}_n(i) \min\{e^{-\gamma \varrho_i}, \Gamma A_{n-1}V^{\gamma}_{n-1}(c)\}}{\sum_{i=0}^{i_{\max}} \hat{p}_n(i) \min\{e^{-\gamma \varrho_i}, \Gamma A_{n-1}V^{\gamma}_{n-1}(c-1)\}} \\ & \geq \Gamma A_{n-1}V^{\gamma}_{n-1}(c-1)\,. \end{aligned}$$

Similarly,

$$\begin{aligned} \Gamma A_n V^{\gamma}_n(c) =& \Gamma A_{n-1}V^{\gamma}_{n-1}(c) \\ & \cdot \frac{\sum_{i=0}^{i_{\max}} \hat{p}_n(i) \min\{1,\ e^{-\gamma \varrho_i}(1/\Gamma A_{n-1}V^{\gamma}_{n-1}(c))\}}{\sum_{i=0}^{i_{\max}} \hat{p}_n(i) \min\{1,\ e^{-\gamma \varrho_i}(1/\Gamma A_{n-1}V^{\gamma}_{n-1}(c-1))\}} \\ & \leq \Gamma A_{n-1}V^{\gamma}_{n-1}(c) \end{aligned}$$

results from using (7.6) and Lemma 7.5 (i) again. Hence,

$$\Gamma A_{n-1}V^{\gamma}_{n-1}(c-1) \leq \Gamma A_n V^{\gamma}_n(c) \leq \Gamma A_{n-1}V^{\gamma}_{n-1}(c)\,.$$

By definition of the protection levels, for $c = y^{\gamma*}_{i-1}(n)$,

$$\exp(-\gamma \varrho_i) > \Gamma A_n V^{\gamma}_n(y^{\gamma*}_{i-1}(n)) \geq \Gamma A_{n-1}V^{\gamma}_{n-1}(y^{\gamma*}_{i-1}(n) - 1)\,,$$

which implies $y_{i-1}^{\gamma *}(n-1) \geq y_{i-1}^{\gamma *}(n) - 1$. Analogously, for $c = y_{i-1}^{\gamma *}(n-1)$,

$$\exp(-\gamma \varrho_i) > \Gamma A_{n-1} V_{n-1}^{\gamma}(y_{i-1}^{\gamma *}(n-1)) \geq \Gamma A_n V_n^{\gamma}(y_{i-1}^{\gamma *}(n-1)) ,$$

which implies $y_{i-1}^{\gamma *}(n) \geq y_{i-1}^{\gamma *}(n-1)$.

Let us now turn to assertion (i). That $y_{i-1}^{\gamma *}(n)$ is increasing in i follows from Lemma 7.2 or directly from the definition of $y_{i-1}^{\gamma *}(n)$ in Theorem 7.1 together with Lemma 7.5 (i) and $\varrho_i \geq \varrho_{i+1}$.

The protection level $y_0^{\gamma *}(n)$ equals 0 because

$$\Gamma A_n G_n^{\gamma}(c) \geq \exp(-\gamma \varrho_1)$$

for all $c > 0$. The latter inequality follows by induction. For $n = 0$ and $c > 0$, the left-hand side is 1 and therefore larger than $\exp(-\gamma \varrho_1)$. Now assume that this is true for some n. Replacing ϱ_i for ϱ_1 in the calculation of $G_{n+1}^{\gamma}(c)$ and substituting $G_{n+1}^{\gamma}(c-1)$ by $G_n^{\gamma}(c-1)$ (making the divisor in $\Gamma A_{n+1} G_{n+1}^{\gamma}(c)$ larger by Lemma 7.3) yields

$$\Gamma A_{n+1} G_{n+1}^{\gamma}(c) \geq \min\left\{\exp(-\gamma \varrho_1), \Gamma A_n G_n^{\gamma}(c)\right\} = \exp(-\gamma \varrho_1) ,$$

where the last equality follows from the induction hypothesis. Consequently, the left-hand side is larger than $\exp(-\gamma \varrho_1)$ for $c \geq 1$. From Lemma 7.4, we know that $\Gamma A_{n+1} G_n^{\gamma}(0) = \exp(-\gamma \bar{\varrho}) < \exp(-\gamma \varrho_1)$ for all $n = 1, \ldots, N$. Thus, (i) also holds, completing the proof. □

Finally, we can conclude that in the dynamic model, all structural properties of the optimal policy that are well-known in the risk-neutral case also hold for a policy maximizing exponential atemporal utility with $\gamma > 0$.

Indeed, exponential utility maximization can be seen as an extension of the risk-neutral objective in the following sense: For $\gamma \to 0$, the certainty equivalent of $V_n^{\gamma}(c)$ converges to $V_n(c)$. If there is a policy that is optimal for all sufficiently small values of γ, this policy is the optimal policy for a risk-neutral decision-maker.

Proposition 7.2. *Protection levels $y_{i-1}^{\gamma *}(n)$ that are γ-optimal for all $\gamma \in (0, \gamma_0)$ are expected revenue maximizing protection levels $y_{i-1}^{*}(n)$.*

Proof. First, note that for finite random variables W_i and values $\hat{p}_i$ with $\sum_i \hat{p}(i) = 1$, $\hat{p}(i) \geq 0$ for all i, a Taylor expansion around $\gamma = 0$ yields

$$\begin{aligned} -\frac{1}{\gamma} \ln\left(\sum_i \hat{p}(i) \mathbb{E}[\exp(-\gamma W_i)]\right) &= -\frac{1}{\gamma} \ln\left(1 - \gamma \sum_i \hat{p}(i) \mathbb{E}[W_i] + O(\gamma^2)\right) \\ &= \sum_i \hat{p}(i) \mathbb{E}[W_i] + O(\gamma) , \end{aligned} \tag{7.8}$$

similar to (3.6).

We can therefore conclude that for a fixed policy π, the certainty equivalent of the maximum expected exponential utility of total reward before the request type is known converges to the maximum expected total reward before the request type is known. Since this holds for all policies, it also holds for the maximum of the finite set of all policies consisting of decision rules of protection level type and protection levels in $\{0,\ldots,n-1\}$, which we denote by $\mathfrak{F}_y^N$, i.e.

$$\begin{aligned}-\frac{1}{\gamma}\ln(-A_nV_n^\gamma(c)) &= \max_{\pi\in\mathfrak{F}_y^N} -\frac{1}{\gamma}\ln\left(\sum_{i=0}^{i_{\max}}\hat{p}_n(i)\mathbb{E}_\pi[\exp(-\gamma R_\pi)|X_n=(c,i)]\right)\\ &\to \max_{\pi\in\mathfrak{F}_y^N}\sum_{i=0}^{i_{\max}}\hat{p}_n(i)\mathbb{E}_\pi[R_\pi|X_n=(c,i)] = A_nV_n(c)\,,\end{aligned}$$

cf. Theorems 7.1 and 5.1.

Now rearrange the γ-optimal protection levels to read

$$y_{i-1}^{\gamma*}(n) = \max\left\{c\in\{0,\ldots,n-1\}\ :\ \varrho_i < -\frac{1}{\gamma}\ln\left(\frac{A_{n-1}V_{n-1}^\gamma(c)}{A_{n-1}V_{n-1}^\gamma(c-1)}\right)\right\}.$$

After another rearrangement of the right-hand side of the inequality and using $-\frac{1}{\gamma}\ln(-A_nV_n^\gamma(c)) \to A_nV_n(c)$, it is clear that for $\gamma\to 0$,

$$-\frac{1}{\gamma}\ln\left(-A_{n-1}V_{n-1}^\gamma(c)\right) + \frac{1}{\gamma}\ln\left(-A_{n-1}V_{n-1}^\gamma(c-1)\right) \to \Delta A_{n-1}V_{n-1}(c)\,.$$

Thus, the right side converges to $\Delta A_{n-1}V_{n-1}(c)$. We obtain the protection levels of the optimal, expected revenue maximizing model. □

7.1.4 Numerical Example in the Case of an Exponential Utility Function

To illustrate our structural results, we take up the four-class example by Lee and Hersh (1993) again. The parameters are given in Sect. 5.1.3.

Figures 7.3 and 7.4 show the time-dependent values $y_{i-1}^{\gamma*}(n)$ of the γ-optimal protection levels for $\gamma = 0.001$ and $\gamma = 0.005$. The values of $y_1^{\gamma*}$, $y_2^{\gamma*}$, and $y_3^{\gamma*}$ are indicated in black, gray, and white, respectively. In accord with Theorem 7.1, the protection levels are increasing in i and n by increments of a height of 1.

The fact that γ-optimal protection levels are decreasing in the coefficient of risk-aversion and are smaller than under risk-neutrality is not surprising. A more risk-averse decision-maker values the chance of making revenue from reserving a seat less than a risk-neutral decision-maker. Thus, protection levels are smaller.

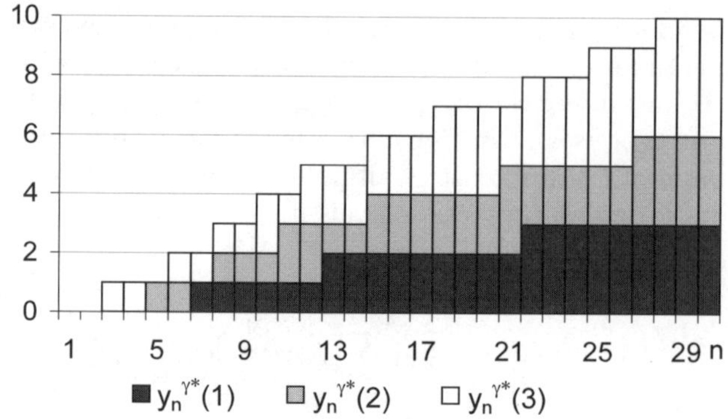

Fig. 7.3. Protection levels of a γ-optimal policy with $\gamma = 0.001$.

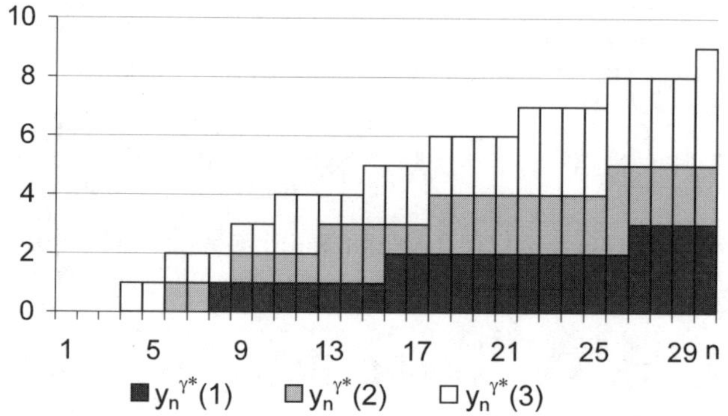

Fig. 7.4. Protection levels of a γ-optimal policy with $\gamma = 0.005$.

7.2 The Static Model

The maximum expected utility of total revenue observing a demand of d in class i and given c seats as well as a current wealth of w in the static model is

$$V^{\text{atmp}*}(c,d,w) = \max_{\pi \in \Pi} \mathbb{E}_\pi \left[u\left(R_\pi\right) \mid (X_N, W_N) = (c,d,w) \right] ,$$

where u is the increasing utility function, and $W_n \in \mathfrak{W}$ is the additional wealth state as defined in Sect. 3.2.2. The total reward (added on the current wealth) is $R_\pi = W_N + \sum_{n=1}^N r_n(X_n, f_n(X_n, W_n)) + V_0(X_N)$. Again, the decision rule might depend on the wealth as well as the system state.

According to the results stated in Sect. 3.2.2, this is the unique solution to the optimality equation

$$V_i^{\text{atmp}}(c,d,w) = \max_{a \in \{0,\ldots,d\}} \left\{ \sum_{d'=0}^{d_{\max}} \hat{p}_{i-1}(d') V_{i-1}^{\text{atmp}}(c-a, d', w + a\varrho_i) \right\} \tag{7.9}$$

with $V_0^{\text{atmp}}(c,d,w) = u(V_0(c,d)+w)$. Every policy $\pi^{\text{atmp}*}$ formed by actions $a = f_n^{\text{atmp}*}(c,d,w)$, each maximizing the right-hand side of the optimality equation, is atmp-optimal, i.e. leads to $V^{\text{atmp}*}$.

Using $\tilde{A}_i V_i^{\text{atmp}}(c,w) = \sum_{d'=0}^{d_{\max}} \hat{p}_i(d') V_i^{\text{atmp}}(c,d',w)$, this should again be reformulated to

$$\tilde{A}_i V_i^{\text{atmp}}(c,w) = \sum_{d=0}^{d_{\max}} \hat{p}_i(d) \max_{a=0,\dots,d} \left\{ \tilde{A}_{i-1} V_{i-1}^{\text{atmp}}(c-a, w+a\varrho_i) \right\} ,$$

with initial value $\tilde{A}_0 V_0^{\text{atmp}}(c,w)$, in order to reduce the state space and computational complexity when applied in practice.

Given only two fare classes, we mentioned before that the static capacity control model reduces to the classic newsvendor problem. Maximization of expected utility in this special case is discussed e.g. in Eeckhoudt et al. (1995), who show that the protection level for class 1 customers decreases for increasing risk-aversion.

7.2.1 Numerical Examples

Again, one can see that given a general increasing utility function, many structural properties known from the risk-neutral model might fail. The first example is another toy example to show that monotonicity of an atmp-optimal policy in c and w is generally not true for all increasing utility functions u. The second is a continuation of the running example based on van Ryzin and McGill (2000). In the case of a logarithmic utility function, the atmp-optimal action is monotone in the remaining capacity in this example. But in contrast to the risk-neutral case, capacity-independent protection levels need not exist. Monotonicity in i holds for both examples.

Example 1: Non-Monotonicity in c and w

Imagine a $i_{\max} = 3$ class model with associated prices of $\varrho_1 = 5$, $\varrho_2 = 2$, and $\varrho_3 = 1$. We consider independent customer arrival probabilities of $\hat{p}_i(0) = 0.8$, $\hat{p}_i(1) = \hat{p}_i(2) = 0.05$, and $\hat{p}_i(3) = 0.1$ for booking classes $i = 1, 2$.

First let us assume that the decision-maker's preferences can be represented by a utility function with an aspiration level of $\beta = 6$. Now consider the following scenario: There are $d = 6$ requests from class $i = 3$ customers, and the current wealth is $w = 0$.

Given only one piece of capacity, $c = 1$, the aspiration level is not reachable, so the atmp-optimal decision is arbitrary. Given $c = 2$, the decision-maker should accept one of the requests. Given $c = 3$, rejecting all requests is atmp-optimal. Given $c = 4$, the decision-maker should accept two of the requests, and for $c = 5$, four of them should be accepted. For values of $c \geq 6$, the aspiration level can be met directly by accepting all six requests, which is the optimal action to take. Clearly, the actions are not monotone in the remaining capacity c. The reason for the strange behavior at $c = 3$ is the following:

With a capacity of $c = 2$, the only way of earning 6 or more is to sell one seat to a class 1 and one to any other fare class customer. If there is class 3 demand, the decision-maker should consequently accept one request and hope for a future class 1 customer.

With a capacity of $c = 3$, there are more ways of earning 6 or more. The decision of whether to accept one class 3 request can be reduced to a comparison of a "there will be three or more class 2 requests and no class 1 request" scenario with "there will be exactly one class 1 request". Since the first scenario is the more probable one, it is better to deny all customer class 3 requests. With $c = 4$, the scenario comparisons again favor accepting up to two customer requests.

Given an initial wealth of $w = 1$ (which is the same as reducing the aspiration level to $\beta = 5$), the decision-maker prefers to reject all requests given $c = 1$ and $c = 2$. For $c = 3$, he should accept one request. Thus, the atmp-optimal action is also not monotone in w. The reason is again attributable to combinatorial effects.

Example 2: The Example given in van Ryzin and McGill (2000) (Continued)

Let us turn to the example based on the parameters given in van Ryzin and McGill (2000) as stated in Sect. 5.2.3. This time we assume that the decision-maker maximizes expected atemporal utility from selling $C = 100$ seats. If the utility function has an aspiration level of e.g. 60000, the decision problem again turns into a pure combinatorial problem. Given the more well-behaved logarithmic utility function $u(w) = \ln(w+1)$, there is an atmp-optimal policy for this example that can be described in terms of atmp-optimal controls $y_1^{\text{atmp}*}(c,w) \leq y_2^{\text{atmp}*}(c,w) \leq y_3^{\text{atmp}*}(c,w)$, so that the atmp-optimal action is $a_i(c,d,w) = \max\{0, c - y_i^{\text{atmp}*}(c,w)\}$. Note that these controls depend on the remaining capacity c and current wealth w in addition to the booking class i. Given a wealth of $w = 0$ and $w = 50000$, the controls $y_1^{\text{atmp}*}(c,w)$ and $y_2^{\text{atmp}*}(c,w)$ are depicted in Figs. 7.5 and 7.6. As usual, the controls of classes 1 and 2 are shaded black and gray, respectively. $y_3(c,0)$ and $y_3(c,50000)$ are greater than 100 and are therefore not shown.

Using the same reasoning as in Example 2 of Sect. 7.1.2 concerning atmp-optimal policies for the dynamic model, it is not surprising that the protection levels are increasing in wealth w. The controls are increasing in c, yet they never increase by more than 1, so that the atmp-optimal action is increasing in the remaining capacity.

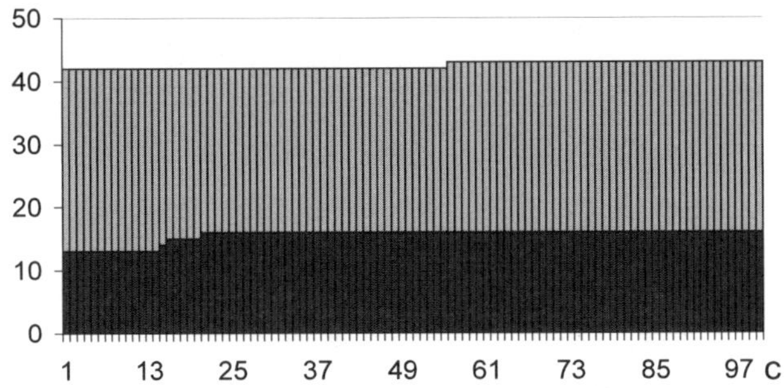

Fig. 7.5. Atmp-optimal controls given a decision-maker with logarithmic utility function at current wealth $w = 0$.

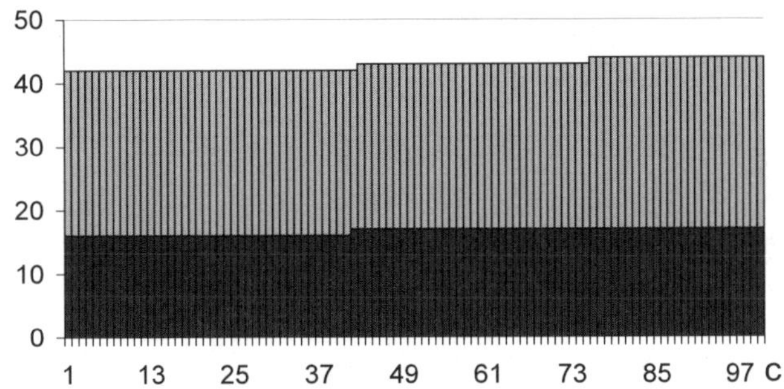

Fig. 7.6. Atmp-optimal controls given a decision-maker with logarithmic utility function at current wealth $w = 50000$.

7.2.2 Structural Results in the Case of an Exponential Utility Function

In the following, we concentrate on a risk-averse decision-maker who seeks to maximize the expected atemporal exponential utility of the total reward. Our aim is therefore to determine a policy $\pi^{\gamma *} = (f^{\gamma *}_{i_{\max}}, f^{\gamma *}_{i_{\max}-1}, \dots, f^{\gamma *}_1)$, called γ-optimal, which realizes

$$V^{\gamma *}(c,d) = \max_{\pi \in \Pi} \mathbb{E}_\pi \left[- \exp\left(-\gamma R_\pi\right) \mid X_{i_{\max}} = (c,d) \right] .$$

The corresponding optimality equation reads

$$V^{\gamma}_i(c,d) = \max_{a \in \{0,\dots,d\}} \left\{ \exp(-\gamma a \varrho_i) \sum_{d'=0}^{d_{\max}} \hat{p}_{i-1}(d') V^{\gamma}_{i-1}(c-a,d') \right\} , \qquad (7.10)$$

where $V^{\gamma}_0(c,d) = -\exp(-\gamma V_0(c,d))$. Again, $V^{\gamma *} \equiv V^{\gamma}_{i_{\max}}$, and every policy $\pi^{\gamma *}$ formed by actions $f^{\gamma *}_i(c,d)$, each maximizing the right-hand side of (7.10), is γ-optimal.

As in the dynamic model, it turns out to be be more convenient to work with $G_i^\gamma = -V_i^\gamma$, which is the unique solution of

$$G_i^\gamma(c,d) = \min_{a\in\{0,\dots,d\}} \{\exp(-\gamma a\varrho_i)A_{i-1}G_{i-1}^\gamma(c-a)\} \tag{7.11}$$

$$= A_{i-1}G_{i-1}^\gamma(c) \cdot \min_{a=0,\dots,d} \prod_{a'=1}^{a} \exp(-\gamma\varrho_i)\frac{A_{i-1}G_{i-1}^\gamma(c-a')}{A_{i-1}G_{i-1}^\gamma(c-a'+1)} \tag{7.12}$$

with an initial value $G_0^\gamma = -V_0^\gamma$. This transformation preserves the γ-optimality of a policy.

Of course, a γ-optimal policy accepts all requests if the number of remaining seats is higher than the maximum number of current and future requests possible. We choose $\bar{\varrho}$ to ensure that overbooking is never advisable.

Lemma 7.6. *For $\gamma > 0$, $i \in \{1,\dots,i_{max}\}$, and $d \in \{0,\dots,d_{max}\}$ we have*

(i) $G_i^\gamma(c,d) = \exp(-\gamma\varrho_i d)A_{i-1}G_{i-1}^\gamma(c-1)$
$= \exp(-\gamma\varrho_i d)\prod_{i'=1}^{i-1}\sum_{d'=0}^{d_{max}}\hat{p}_{i'}(d')\exp(-\gamma\varrho_{i'}d')$, $c \geq id_{max}$.

(ii) $G_i^\gamma(c,d) = A_{n-i}G_{i-1}^\gamma(c) = \exp(-\gamma\bar{\varrho}c)$, $\quad c \leq 0$.

Proof. (i) and (ii) follow by induction on i using the inequalities $\exp(-\gamma\varrho_i) \leq 1 \leq \exp(\gamma(\bar{\varrho}-\varrho_i))$.

Both assertions are certainly true for $n = 1$, since

$$G_1^\gamma(c,d) = \min_{a=0,\dots,d}\{\exp(-\gamma\varrho_1 a)\exp(-\gamma V_0(c-a,0))\} .$$

For $c \leq 0$, the terminal reward is $V_0(c-a,0) = (c-a)\bar{\varrho}$. It follows that

$$\exp(-\gamma\varrho_1 a)\exp(-\gamma V_0(c-a,0)) = \exp(-\gamma\bar{\varrho}c)\exp(\gamma a(\bar{\varrho}-\varrho_1)) ,$$

where the last term increases in a, since $\bar{\varrho} > \varrho_1$. Thus, the term is minimal for $a = 0$ satisfying (i). For $c \geq d_{\max}$, every action results in a non-negative capacity at $i = 0$ with a terminal reward of 0. Consequently, the only term of interest is $\exp(-\gamma\varrho_1 a)$, which is decreasing in a. Therefore, the minimum is attained for $a = d$ in accordance with (ii).

Now assume that (i) and (ii) are true for some i.

$$G_{i+1}^\gamma(c,d) = \min_{a=0,\dots,d}\left\{\exp(-\gamma\varrho_{i+1}a)\sum_{d'=0}^{d_{\max}}\hat{p}_i(d')G_i^\gamma(c-a,d')\right\} .$$

Using the induction hypothesis for $c \leq 0$, this again reduces to

$$G_{i+1}^\gamma(c,d) = \min_{a=0,\dots,d}\{\exp(-\gamma\bar{\varrho}c)\exp(\gamma a(\bar{\varrho}-\varrho_{i+1}))\} ,$$

where the minimum is attained for $a = 0$, resulting in (i). For $c \geq (i+1)d_{\max}$, every action causes a transition to a state $c - a \geq id_{\max}$. Using the induction hypothesis, this yields

$$G_{i+1}^{\gamma}(c,d) = \min_{a=0,\ldots,d} \left\{ \exp(-\gamma \varrho_{i+1} a) \prod_{i'=1}^{i} \sum_{d'=0}^{d_{\max}} \hat{p}_{i'}(d') \exp(-\gamma \varrho_{i'} d') \right\} .$$

Again, this is minimal for $a = d$, proving (ii). □

In addition, the following lemma is needed to show that there is a γ-optimal policy of protection level type.

Lemma 7.7. *For $\gamma > 0$, $d \in \mathbb{N}_0$, and $i \in \{1, \ldots, i_{max}\}$, it holds that*

(i) $A_{i-1} G_{i-1}^{\gamma}(c)$ is log-convex and decreasing in c.
(ii) $G_i^{\gamma}(c,d)$ is log-convex and decreasing in c.

Proof. The assertions follow essentially from the same arguments as given in the proof of Lemma 7.5 for fixed $\gamma > 0$. Setting

$$g_{i-1}(c) = A_{i-1} G_{i-1}^{\gamma}(c) \quad c \in \mathbb{Z}$$

for $i = 1, \ldots, i_{\max}$, we can restate the optimality equation (7.11) to read

$$\ln G_i^{\gamma}(c,d) = \min_{a \in \{0,\ldots,d\}} \{-\gamma \varrho_i a + \ln g_{i-1}(c-a)\} . \tag{7.13}$$

Let $i = 1$. Then since $-\gamma V_0(\cdot, d)$ is decreasing and convex, and using the closure of log-convex functions with respect to convex combinations, we have log-convexity of g_0. In addition, we know that g_0 is decreasing.

Since the minimum of decreasing functions is decreasing, $\ln G_1^{\gamma}(\cdot, d)$, and thus $G_1^{\gamma}(\cdot, d)$ is decreasing. By applying Lemma 1 in Stidham (1978) (see Appendix B) to (7.11), we can additionally conclude that $\ln G_1^{\gamma}(\cdot, d)$ is convex, $d = 0, \ldots, d_{\max}$. Consequently, $G_1^{\gamma}(\cdot, d)$ is log-convex.

Now suppose $G_{i-1}^{\gamma}(\cdot, d)$ is decreasing and log-convex for some $1 \leq i < i_{\max}$. Then $g_{i-1}(c)$ is a convex combination of decreasing and log-convex functions. Hence, $g_{i-1}(c)$ is decreasing and log-convex. By applying Stidham's lemma (1978) to (7.11), we finally get the desired log-convexity of $G_i^{\gamma}(\cdot, d)$, $d = 0, \ldots, d_{\max}$. In addition, we can conclude that $G_i^{\gamma}(\cdot, d)$ is decreasing. □

We are now in a position to prove the main result of this model.

Theorem 7.2. *For $\gamma > 0$, there exists a γ-optimal policy $\pi^{\gamma *} = (f_{i_{max}}^{\gamma *}, f_{i_{max}-1}^{\gamma *}, \ldots, f_1^{\gamma *})$ such that*

$$f_i^{\gamma *}(c,d) = \begin{cases} \min\{d, c - y_{i-1}^{\gamma *}\} & c > y_{i-1}^{\gamma *} \\ 0 & c \leq y_{i-1}^{\gamma *} , \end{cases}$$

with

$$y_{i-1}^{\gamma *} = \max \left\{ c \in \{0, \ldots, (i-1)d_{max}\} : \exp(-\gamma \varrho_i) > \Gamma A_{i-1} G_{i-1}^{\gamma}(c) \right\} .$$

Proof. Fix $\gamma > 0$. According to Lemma 7.7 (i), $A_{i-1}G^{\gamma}_{i-1}(c)$ is log-convex and decreasing in c. Hence, $A_{i-1}G^{\gamma}_{i-1}(c)$ is a positive function and the ratio

$$\Gamma A_{i-1}G^{\gamma}_{i-1}(c) = \frac{A_{i-1}G^{\gamma}_{i-1}(c)}{A_{i-1}G^{\gamma}_{i-1}(c-1)} \leq 1$$

is increasing in c.

From (7.12), it follows that the largest action a with

$$\exp(-\gamma\varrho_i) \leq \Gamma A_{i-1}G^{\gamma}_{i-1}(c-a+1)$$

is γ-optimal. Together with Lemma 7.6, we can conclude that the γ-optimal policy is of protection level type and that the constants $y^{\gamma*}_{i-1}$ as defined above are γ-optimal protection levels for $i = 0, 1, \ldots, i_{\max} - 1$. □

Proposition 7.3. *γ-optimal protection levels are increasing in i, i.e.*

$$0 = y^{\gamma*}_0 \leq y^{\gamma*}_1 \leq \ldots \leq y^{\gamma*}_{i_{max}-1} \ .$$

Proof. Fix $\gamma > 0$. $y^{\gamma*}_0 = 0$ follows directly from the definition of $V_0(c,d)$ and the choice of $\bar{\varrho} > \varrho_1$.

Now, let us turn to the monotonicity in i. As a first step, we show that $\Gamma A_{i-1}G^{\gamma}_{i-1}(c)$ is decreasing in i, i.e.

$$\Gamma A_{i-1}G^{\gamma}_{i-1}(c) \geq \frac{\sum_{d=0}^{d_{\max}} \hat{p}_i(d) \min_{a=0,\ldots,d}\{\exp(-\gamma\varrho_i a)A_{i-1}G^{\gamma}_{i-1}(c-a)\}}{\sum_{d=0}^{d_{\max}} \hat{p}_i(d) \min_{a=0,\ldots,d}\{\exp(-\gamma\varrho_i a)A_{i-1}G^{\gamma}_{i-1}(c-1-a)\}}$$

for all $i = 1, \ldots, i_{\max}$. This condition is equivalent to

$$\begin{aligned}
&\sum_{d=0}^{d_{\max}} \hat{p}_i(d) \min_{a=0,\ldots,d} \left\{\exp(-\gamma\varrho_i a)\frac{A_{i-1}G^{\gamma}_{i-1}(c-1-a)}{A_{i-1}G^{\gamma}_{i-1}(c-1)}\right\} \\
&\leq \sum_{d=0}^{d_{\max}} \hat{p}_i(d) \min_{a=0,\ldots,d} \left\{\exp(-\gamma\varrho_i a)\frac{A_{i-1}G^{\gamma}_{i-1}(c-a)}{A_{i-1}G^{\gamma}_{i-1}(c)}\right\} ,
\end{aligned}$$

which is true, since

$$\frac{A_{i-1}G^{\gamma}_{i-1}(c-a)}{A_{i-1}G^{\gamma}_{i-1}(c)} = \prod_{a'=1}^{a} \frac{A_{i-1}G^{\gamma}_{i-1}(c-a')}{A_{i-1}G^{\gamma}_{i-1}(c-a'+1)}$$

is decreasing in c for all $i = 1, \ldots, i_{\max}$, owing to the log-convexity of $A_{i-1}G^{\gamma}_{i-1}(c)$ cf. Lemma 7.7.

Since the booking classes are ordered by fare ϱ_i with $\varrho_{i+1} < \varrho_i$ for all $i = 1, \ldots, i_{\max} - 1$, the term $\exp(-\gamma\varrho_i)$ is increasing in i. Hence, for all $i = 1, \ldots, i_{\max} - 1$,

$$\begin{aligned} y_{i-1}^{\gamma*} &= \max\left\{c \in \{0,\ldots,(i-1)d_{\max}\} : \exp(-\gamma\varrho_i) > \Gamma A_{i-1}G_{i-1}^{\gamma}(c)\right\} \\ &\leq \max\left\{c \in \{0,\ldots,id_{\max}\} : \exp(-\gamma\varrho_{i+1}) > \Gamma A_i G_i^{\gamma}(c)\right\} \\ &= y_i^{\gamma*} \end{aligned}$$

holds, completing the proof. □

Thus, we can conclude that in the static model, there is a γ-optimal policy with structural properties well-known from the risk-neutral case.

As in the dynamic model, we can show that in fact, the risk-sensitive case can be seen as a generalization of the risk-neutral case in the following sense:

Proposition 7.4. *Protection levels $y_{i-1}^{\gamma*}$ that are γ-optimal for all $\gamma \in (0,\gamma_0)$ are expected revenue maximizing protection levels y_{i-1}^{*}.*

Proof. From (7.8), we know that for a fixed policy π, the certainty equivalent of the maximum expected exponential utility of total reward before the number of requests is known converges to the maximum expected total reward before the number of requests is known as γ approaches 0. Since this holds for all policies, it also holds for the maximum

$$-\frac{1}{\gamma}\ln(-A_iV_i^{\gamma}(c)) = \max_{\pi\in\mathfrak{F}_y^{i\max}} -\frac{1}{\gamma}\ln\left(\sum_{d=0}^{d_{\max}} \hat{p}_i(d)\mathbb{E}_{\pi}[\exp(-\gamma R_{\pi})|X_i=(c,d)]\right)$$

$$\rightarrow \max_{\pi\in\mathfrak{F}_y^{i\max}} \sum_{d=0}^{d_{\max}} \hat{p}_i(d)\mathbb{E}_{\pi}[R_{\pi}|X_i=(c,d)] = A_iV_i(c)\;,$$

where $\mathfrak{F}_y$ denotes the finite set of all decision rules of protection level type and protection levels in $\{0,\ldots,id_{\max}\}$ (cf. Theorems 7.2 and 5.2).

If we solve the inequality in the definition of the γ-optimal protection levels for ϱ_i and use $V_i^{\gamma}(c) = -G_i^{\gamma}(c)$, we obtain

$$y_{i-1}^{\gamma*} = \max\left\{c \in \{0,\ldots,id_{\max}\} \;:\; \varrho_i < -\frac{1}{\gamma}\ln\left(\frac{A_{i-1}V_{i-1}^{\gamma}(c)}{A_{i-1}V_{i-1}^{\gamma}(c-1)}\right)\right\}.$$

Rearranging the right-hand side of the inequality and using $-\frac{1}{\gamma}\ln(-A_iV_i^{\gamma}(c)) \rightarrow A_iV_i(c)$ gives

$$-\frac{1}{\gamma}\ln\left(-A_{i-1}V_{i-1}^{\gamma}(c)\right) + \frac{1}{\gamma}\ln\left(-A_{i-1}V_{i-1}^{\gamma}(c-1)\right) \rightarrow \Delta A_{n-1}V_{i-1}(c)$$

for $\gamma \rightarrow 0$. Thus, the right side converges to $\Delta A_{i-1}V_{i-1}(c)$, which yields the protection levels of the optimal, expected revenue maximizing policy. □

7.2.3 Numerical Example in the Case of an Exponential Utility Function

Let us again consider the example data from van Ryzin and McGill (2000) as given in Sect. 5.2.3; this time, however, we assume a decision-maker maximizing expected atemporal exponential utility. Remember that there are 4 fare classes with fare prices of $\varrho_1 = 1050 \geq \varrho_2 = 567 \geq \varrho_3 = 527 \geq \varrho_4 = 350$. The optimal expected revenue maximizing protection levels are $y_3^* = 133$, $y_2^* = 44$, and $y_1^* = 17$, with expected revenue of about 60038.

Now consider e.g. a decision-maker maximizing expected atemporal exponential utility with $\gamma = 0.0001$. The γ-optimal protection levels are $y_3^{0.0001*} = 118$, $y_2^{0.0001*} = 39$, and $y_1^{0.0001*} = 15$, which result in an expected revenue of 59906. Of course, these protection levels are increasing in i in accordance with Proposition 7.3.

To analyze the impact of γ-optimal protection levels in more detail, the programming language Java was used to code the optimization as well as a simulation of the resulting request arrival and booking process given various γ-optimal protection levels. In addition to the expected revenue maximizing case, 150 γ-optimal policies were considered in total. 50 γ-optimal policies were generated by increasing γ in steps of 0.00001 starting from 0; the step size was then increased to 0.0001 for another 50 policies. Finally, the step size was set at 0.0005. For every γ-optimal policy, 500000 trials of the booking process were run to ensure statistically significant results. The results are based on a simulation study in Barz (2006).

Let us consider the effect of γ in this example in more detail. Intuitively, increasing risk-aversion causes the decision-maker to increasingly prefer the sure revenue of low-fare classes to the chance of making more revenue from future fare classes. Thus, protection levels decrease for increasingly risk-averse decision-makers as depicted in Fig. 7.7. In this figure, the values of the pro-

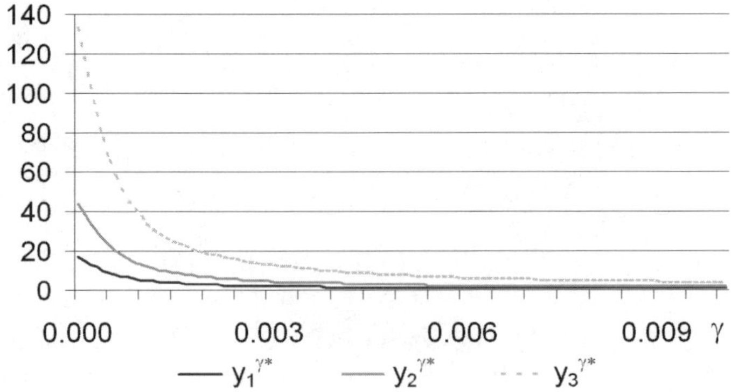

Fig. 7.7. Protection levels of a γ-optimal policy.

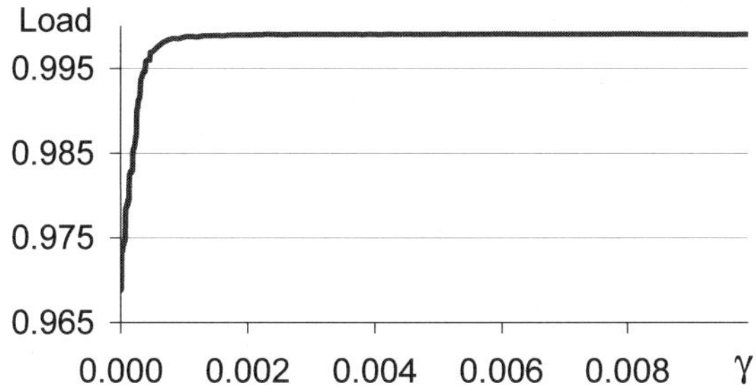

Fig. 7.8. Simulated average load factor given a γ-optimal policy.

tection levels $y_1^{\gamma*}$, $y_2^{\gamma*}$, and $y_3^{\gamma*}$ are shaded black, dark gray, and light gray, respectively.

For small values of γ, the protection levels approach (and finally equal) the levels of the expected revenue maximizing policy, which is in line with Proposition 7.4. For higher values of γ, the focus is increasingly on the worst-case scenario. Thus, protection levels converge to 0, which results in a simple first-come-first-served policy. As a result, the expected load factor increases in γ, which can be seen in Fig. 7.8.

What is the effect of γ on the distribution of R_π, the total revenue gained? It is clear that the largest expected revenue is gained by applying the risk-neutral approach. The expected revenue from applying a γ-optimal policy decreases in γ as can be seen in Fig. 7.9. This makes sense, since the higher the risk-aversion, the more one is willing to forego possible future gains in return for certain revenue. On the other hand, the variance of the revenue gained under the γ-optimal strategy decreases for small values of γ even faster, as can be seen in Fig. 7.10. For $\gamma = 0.0001$, a nearly 18% reduction of standard deviation is dispensed with a revenue loss of only 0.2% compared to the risk-neutral case. (This decrease for small values of γ is in line with the results of

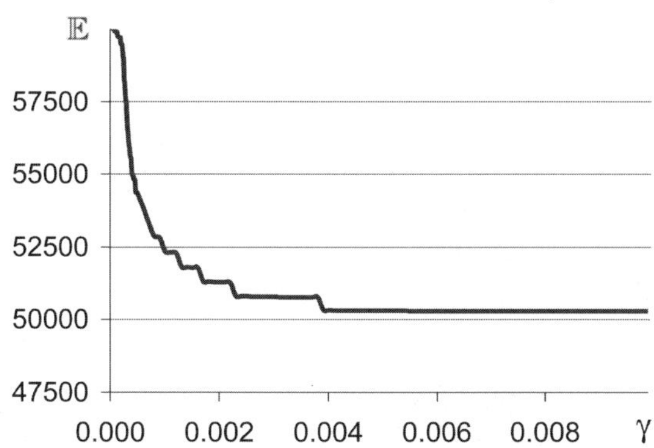

Fig. 7.9. Expected total revenue when applying a γ-optimal policy.

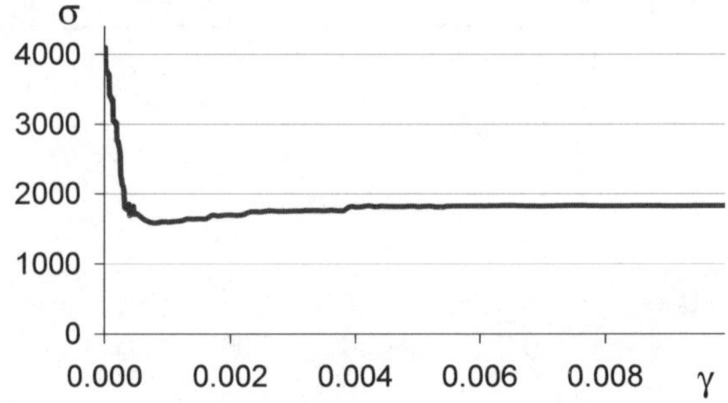

Fig. 7.10. Simulated standard deviation when applying a γ-optimal policy.

the Taylor expansion around $\gamma = 0$; cf. (3.6).) Given these values, it seems plausible that some companies might prefer a non-maximum expected revenue with lower variance to the standard expected revenue maximizing solution gained by traditional revenue management systems.

The zigzag of the standard deviation in the area of $\gamma = 0.0004$ is attributable to the structure of the example data and to effects that arise from jumps in the protection levels. In this region, the demand from classes 2 and 3 suffices to fill their protected seats with a probability of almost 1 and generates a revenue of roughly 550 per seat. Protecting one seat less for those customers (and leaving $y_1^{\gamma*}$ constant) thus turns an almost certain revenue of roughly 550 into a lottery over this value, if less than $C - y_3^{\gamma*}$ class 4 customers arrive, and $\varrho_4 = 350$, else. Hence, the variance may increase. On the other hand, owing to the high class 1 fare price, $y_1^{\gamma*}$ is relatively high. Since the probability that class 1 customers fill all their protected seats is only about 11%, lowering the protection level $y_1^{\gamma*}$ reduces the variance. Consequently, decreasing protection levels (caused by increasing γ) can lead to both higher and lower revenue variance in this area.

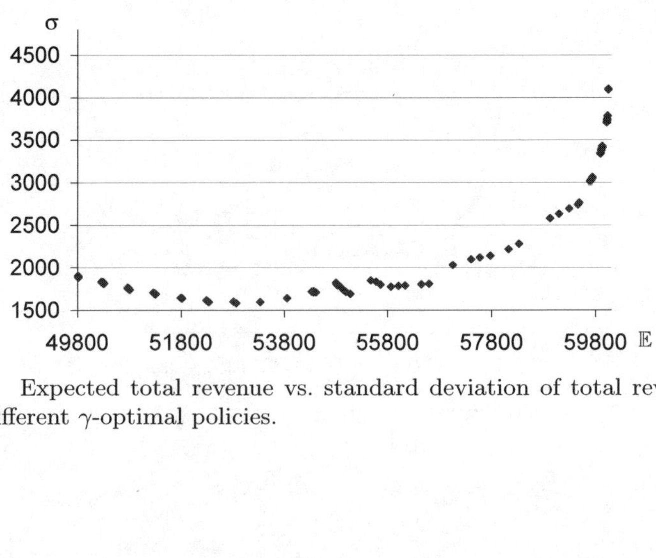

Fig. 7.11. Expected total revenue vs. standard deviation of total revenue when applying different γ-optimal policies.

For high values of γ, a decision-maker maximizing expected atemporal exponential utility focuses on the control of the worst case. Therefore, protection levels tend to 0, and the system behavior converges to the behavior of the uncontrolled process.

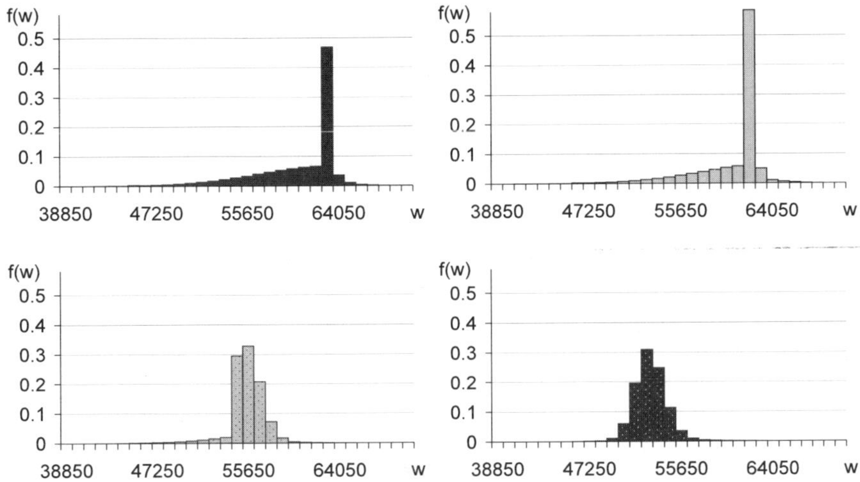

Fig. 7.12. Relative frequencies of total revenue for 500000 simulation runs when applying an optimal (risk-neutral) policy and applying a γ-optimal policy with $\gamma =$ 0.0001, 0.0004, and 0.01 (from upper left to lower right).

Figure 7.11, a plot of the expected revenue vs. the standard deviation for different levels of γ, illustrates this effect from a different perspective. Increasing risk-aversion results in a reduction of the standard deviation at first. For higher values of γ and fixed $y_1^{\gamma *}$, the variance can even increase, but it decreases as soon as $y_1^{\gamma *}$ decreases. Finally, for high risk-aversion, when $y_1^{\gamma *}$ is extremely low, the variance increases even if $y_1^{\gamma *}$ decreases.

Figure 7.12 depicts simulated relative frequencies of total revenue applying an optimal expected revenue maximizing policy and γ-optimal policies with $\gamma = 0.0001$, 0.0004, and 0.01. A comparison between the upper two histograms, the risk-neutral case vs. $\gamma = 0.0001$, illustrates the trade-off between the variance and expectation mentioned above. For a low coefficient of absolute risk-aversion, the modal class shifts to a smaller level, and the relative frequency of total revenue within this class increases. The frequencies of extremely low and high total revenue values decrease. For higher values of γ it can be seen that the probability of reaching low revenue levels further decreases in γ at the cost of reaching extremely high revenue levels.

7.2.4 Possible Extensions of the EMSR Heuristics

We have shown that there is a γ-optimal policy that can be described in terms of protection levels. In particular, these protection levels are independent of the current wealth. Due to this parallel to the risk-neutral case and the popularity of the EMSR heuristics in practice, one might consider a heuristic approach similar to the EMSR heuristics for solving the static model. To tackle this task in more detail, we first revise the two-class model given a risk-averse decision-maker maximizing atemporal exponential utility before we come to the EMSR extensions.

The Two-Class Model

The decision of how many of seats y to protect for class 1 demand is equivalent to choosing one of several lotteries, where a lottery represents the total revenue earned given y.

In the two-class model, the expected exponential utility of revenue given y protected seats (for $d_2 \geq C - y + 1$) is

$$\begin{aligned} &\sum_{d=0}^{y-1} \hat{p}_1(d) u_\gamma \left(\varrho_2(C-y) + d\varrho_1\right) \\ &+ P(D_1 \geq y) u_\gamma \left(\varrho_2(C-y) + y\varrho_1\right) . \end{aligned}$$

If only $y - 1$ are protected, the expected exponential utility of revenue is

$$\begin{aligned} &\sum_{d=0}^{y-1} \hat{p}_1(d) u_\gamma \left(\varrho_2(C-y+1) + d\varrho_1\right) \\ &+ P(D_1 \geq y) u_\gamma \left(\varrho_2(C-y+1) + (y-1)\varrho_1\right) . \end{aligned}$$

Solving for $u_\gamma(\varrho_2)$, we obtain that given a demand $d_2 \geq C-y+1$, the decision-maker would strictly prefer the lottery protecting more seats if and only if

$$-\frac{\sum_{d=0}^{y-1} \hat{p}_1(d) u_\gamma (d\varrho_1) + P(D_1 \geq y) u_\gamma (y\varrho_1)}{\sum_{d=0}^{y-1} \hat{p}_1(d) u_\gamma (d\varrho_1) + P(D_1 \geq y) u_\gamma ((y-1)\varrho_1)} > u_\gamma(\varrho_2) .$$

The equivalent condition in terms of the fare ϱ_2 is

$$\begin{aligned} &-\frac{1}{\gamma} \ln \left[-\sum_{d=0}^{y-1} \hat{p}_1(d) u_\gamma (d\varrho_1) - P(D_1 \geq y) u_\gamma (y\varrho_1) \right] \\ &+\frac{1}{\gamma} \ln \left[-\sum_{d=0}^{y-1} \hat{p}_1(d) u_\gamma (d\varrho_1) - P(D_1 \geq y) u_\gamma ((y-1)\varrho_1) \right] > \varrho_2 . \end{aligned} \tag{7.14}$$

The left-hand side of inequality (7.14) is decreasing in y and can take values in $(0, \varrho_1)$. A decision-maker with constant absolute risk-aversion of γ prefers the lottery with the highest y that fulfills (7.14). If the random revenue gained from class 1 given y remaining seats is seen as a lottery, the two summands on the left-hand side of (7.14) are the certainty equivalents of this lottery given y and $y-1$ seats. The difference is the increase in the certainty equivalent attributable to the yth seat.

Now we can introduce the marginal seat certainty equivalent of seat y (and class i) to be

$$MSCE_i(y) = -\frac{1}{\gamma}\ln\left[-\sum_{d=0}^{y-1}\hat{p}_i(d)u_\gamma(d\varrho_i) - P(D_i \geq y)u_\gamma(y\varrho_i)\right] + \frac{1}{\gamma}\ln\left[-\sum_{d=0}^{y-1}\hat{p}_i(d)u_\gamma(d\varrho_i) - P(D_i \geq y)u_\gamma((y-1)\varrho_1)\right] .$$

Then we know from (7.14) that the γ-optimal protection level, i.e. number of protected seats, $y_1^{\gamma *}$ satisfies

$$MSCE_1(y_1^{\gamma *}) > \varrho_2 \text{ and } MSCE_1(y_1^{\gamma *}+1) \leq \varrho_2 .$$

The γ-optimal protection level is the largest value of y for which the marginal seat certainty equivalent of class 1 is higher than the class 2 fare.

As mentioned before, we can conclude from the properties of the corresponding newsvendor problem analyzed in Eeckhoudt et al. (1995) that the γ-optimal protection level decreases in γ. The more risk-averse a decision-maker is, the fewer seats he protects for high-fare customers.

The MSCE Heuristics

Following the idea of EMSR-a, the MSCE-a heuristic can be obtained by replacing the expected marginal seat revenue $EMSR_i(y)$ in (5.8) by the marginal seat certainty equivalent $MSCE_i(y)$. The number of seats to be protected from class i for class j customers then becomes the highest value of y_{i-1}^j with

$$MSCE_j(y_{i-1}^j) > \varrho_i \text{ and } MSCE_j(y_{i-1}^j + 1) \leq \varrho_i .$$

This calculation is repeated for each future class $j = i-1, \ldots, 1$, and individual protection levels are summed up according to (5.9).

The "b" version of the marginal seat certainty equivalent heuristic (MSCE-b) aggregates demands in the same way as EMSR-b. When considering stage i, an artificial class $(i-1)^\circ$ replaces future demands $1, \ldots, i-1$ as described in (5.10).

As for the revenue $\varrho_{(i-1)^\circ}$ of this artificial class, two approximations seem straightforward. As suggested in Barz (2006), one could use the same formula as for EMSR-b given in (5.11), since it is a crude approximation of the future seat value anyway. But one could also adhere more closely to the notion of choosing a value that approximates the expected value of a future request. Following the latter approach, $\varrho_{(i-1)^\circ}$ should equal the certainty equivalent of a lottery over the future request's revenue. If we set the probability for a class j request with fare ϱ_j, $j = 1, \ldots, i-1$, to $\mathbb{E}[D_j] / \sum_{j=1}^{i-1} \mathbb{E}[D_j]$, this yields

$$\varrho_{(i-1)^\circ} = -\frac{1}{\gamma} \ln \left[-\frac{\sum_{j=1}^{i-1} \mathbb{E}[D_j] u_\gamma(\varrho_j)}{\sum_{j=1}^{i-1} \mathbb{E}[D_j]} \right] . \tag{7.15}$$

The protection levels of MSCE-b are then the highest value of $y_{(i-1)^\circ}$, with

$$MSCE_{(i-1)^\circ}(y_{(i-1)^\circ}) > \varrho_i \text{ and } MSCE_{(i-1)^\circ}(y_{(i-1)^\circ} + 1) \leq \varrho_i \, .$$

From the results of Eeckhoudt et al. (1995), we know that these protection levels decrease for increasing coefficient of risk-aversion γ. This follows directly for MSCE-a and MSCE-b, with fares determined by (5.11). When the high-class fare of MSCE-b is determined by (7.15), the effect is even stronger, because the certainty equivalent decreases in γ. Together with Proposition 7.4, we can conclude that the protection levels obtained by the MSCE heuristics converge to their EMSR counterparts for $\gamma \to 0$.

A Numerical Example

Returning to our numerical example, protection levels generated by MSCE-a and MSCE-b for different values of γ are given in Tables 7.1 and 7.2. $\gamma = 0$ represents the risk-neutral case. π^a (π^b) denotes the policy consisting of decision rules with protection levels $y_i^{\gamma a}$ ($y_i^{\gamma b}$) determined by MCSE-a (MSCE-b). For the values stated, there is no difference in the protection levels resulting from the two MSCE-b variants. For given $0 < \gamma < 0.031$, the differences between the values of $y_i^{\gamma a}$ and $y_i^{\gamma b}$ are never greater than 1. If differences arise, (7.15) is superior to the average fare approach in terms of achieved certainty equivalent, although the difference is almost negligible.

Unsurprisingly, the protection levels are decreasing in γ. A simulation study with the above-mentioned parameters using the MSCE heuristics instead of the exact approach yields very similar graphs for the expected revenues, standard deviations, load factors etc. . Of course, certainty equivalents

Table 7.1. Protection levels $y_1^{\gamma a}$, $y_2^{\gamma a}$ and $y_3^{\gamma a}$ determined by MSCE-a.

γ	$y_1^{\gamma a}$	$y_2^{\gamma a}$	$y_3^{\gamma a}$	$\mathbb{E}[R_{\pi^a}]$	$-1/\gamma \ln(-\mathbb{E}[u_\gamma(R_{\pi^a})])$
0	17	40	127	60010	60010
0.0001	15	36	113	59852	59118
0.0002	13	31	99	59244	57593
0.0004	10	24	73	54845	51447
0.01	1	2	4	50288	3033

Table 7.2. Protection levels $y_1^{\gamma b}$, $y_2^{\gamma b}$ and $y_3^{\gamma b}$ determined by MSCE-b.

γ	$y_1^{\gamma b}$	$y_2^{\gamma b}$	$y_3^{\gamma b}$	$\mathbb{E}[R_{\pi^b}]$	$-1/\gamma \ln(-\mathbb{E}[u_\gamma(R_{\pi^b})])$
0	17	51	131	59895	59895
0.0001	15	45	116	59952	59037
0.0002	13	38	100	59569	57638
0.0004	10	26	69	54855	51441
0.01	1	1	3	50272	3033

are smaller, but for $\gamma \geq 0.001$, the relative difference (compared to the exact solution) is less than 10^{-4}. For smaller values of γ, the average relative difference with EMSR-a is about 0.07%, with a maximum of 0.46% at $\gamma \approx 0.0003$. Using EMSR-b, this average is about 0.1% and the maximum relative deviation is about 0.49% in the same region of γ.

Figure 7.13 shows expected total revenue and the standard deviation of total revenue for different γ-optimal policies and policies resulting from MSCE-a and MSCE-b. In the figure, MSCE-b stands for MSCE-b combined with (7.15), whereas MSCE-b' stands for MSCE-b with the average price approach. To highlight γ-optimal policies, they are connected by straight lines. Figure 7.13 shows that in terms of a comparison of expected revenue and standard deviation, the heuristics also perform well at first sight, although they are dominated by the exact solution. For given expected revenue, however, increases in standard deviation of up to 36% can be observed compared to the exact approach.

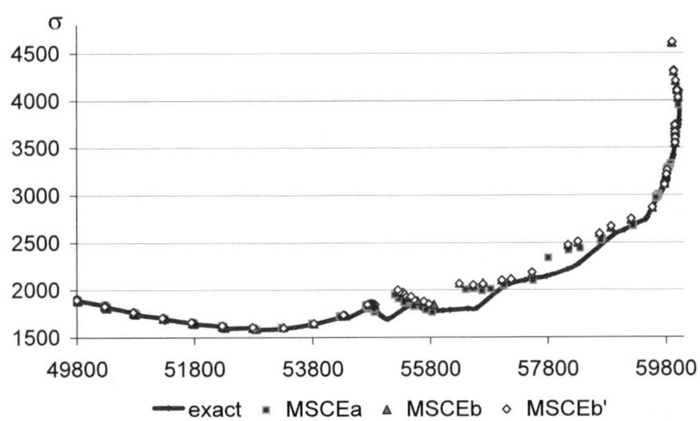

Fig. 7.13. Expected total revenue and standard deviation of total revenue given policies obtained by MSCE-a and MSCE-b for different values of γ.

The EMSU Heuristic

Weatherford (2004) was the first to introduce a risk-averse variant of the EMSR-b heuristic, the so-called expected marginal seat utility (or EMSU) heuristic. Given a risk-averse decision-maker with concave utility function u, he defines the expected marginal seat utility of seat y (at stage i) to be

$$EMSU_i(y) = P(D_i > y)u(\varrho_i) . \qquad (7.16)$$

Using the demand aggregation idea of EMSR-b (see formulas (5.12) and (5.11)[1]), the protection level $y_{(i-1)^\circ}$ for the artificial class $(i-1)^\circ$ from class i is determined by

$$EMSU_{(i-1)^\circ}(y_{(i-1)^\circ}) > u\left(\varrho_i\right) \text{ and } EMSU_{(i-1)^\circ}(y_{(i-1)^\circ}+1) \leq u\left(\varrho_i\right) .$$

Weatherford (2004) introduces EMSU with a general increasing utility function u. In his examples, however, he always makes use of the exponential utility function, $u(x) = 1 - \exp(-\gamma x)$.

The decision-maker's preferences over certain revenue streams are not an explicit issue in Weatherford (2004). Yet even if an exponential utility function is assumed, the preferences implicitly assumed by using EMSU are different from those represented by a maximization of atemporal exponential utility. He motivates and interprets his utility approach for the static model, however, as if it maximized the utility of total revenue. This motivation ends with a statement that the expected marginal utility of making the yth seat available to class i was defined by (7.16). But note that for atemporal (non-linear) utility functions, the utility of selling another seat at fare ϱ_i depends on the current wealth of the decision-maker. (Even in the case of the exponential utility function, the marginal utility depends on the current wealth.) Moreover, we have shown by examples that in the atemporal utility case, there generally need not exist optimal controls of protection level type. If the formulas are analyzed in isolation from the text, it seems as if an expected utility maximizing decision-maker with time-additive utility function is assumed. This would lead to an EMSU-like approach, if the utility function u satisfied $u(0) = 0$ (an assumption that is not explicitly stated in his article) and demand arrived sequentially and independently over time (with at most one arrival within each time period in order of increasing fares). In this case, however, the resulting time periods are rather small and make such time-additive preferences very unlikely. In fact, EMSU seems to be a simple ad-hoc extension of EMSR-b with a utility function.

The Numerical Example (Continued)

The numerical example with the simulation study mentioned above is used to compare the expected atemporal exponential utility maximizing approaches

[1] The fact that Weatherford (2004) uses (5.11) cannot be directly concluded from his article, but from a presentation given in (2003).

to EMSU with exponential utility function for different values of γ. Since the objective function of our approach differs from Weatherford's (2004) EMSU approach, a comparison of utility values or certainty equivalents does not seem appropriate. We therefore stick to the plot of expected revenue vs. standard deviation that is depicted in Fig. 7.14. Of course, the exact expected atemporal exponential utility maximizing policy need not dominate EMSU in terms of expected revenue and standard deviation due to the difference in the objective functions. Yet EMSU has an effect that is similar to the maximization of expected atemporal exponential utility. This can be explained by the transformation of the fares by a concave utility function u with $u(0) = 0$. The transformation moves the fraction $u(\varrho_i)/u(\varrho_{(i-1)^\circ})$ closer to 1 and thus decreases the protection levels for increasing risk-aversion. γ-optimal protection levels decrease for increasing γ, too.

A decision-maker might only be willing to forgo a little revenue for a reduction in variance. Hence, extremely low values of γ, with the corresponding trade-off between expectation and variance, are of special interest for implementations. Figure 7.15 shows a comparison of the MSCE heuristics with EMSU for policies achieving an expected revenue of more than 59000. In this region, EMSU results in policies with a standard deviation of revenue up to one third higher than those generated by MSCE policies with the same or similar expected revenue. MSCE-b outperforms EMSU for standard deviations above 2800 and expected revenues above 59500, respectively. For given lower values of variance, the performance is more mixed due to the difference in the underlying preferences.

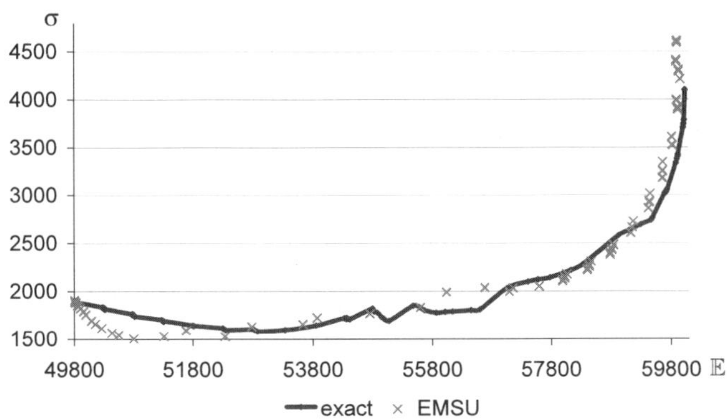

Fig. 7.14. Expected total revenue and standard deviation of total revenue given policies obtained by EMSU and γ-optimal policies.

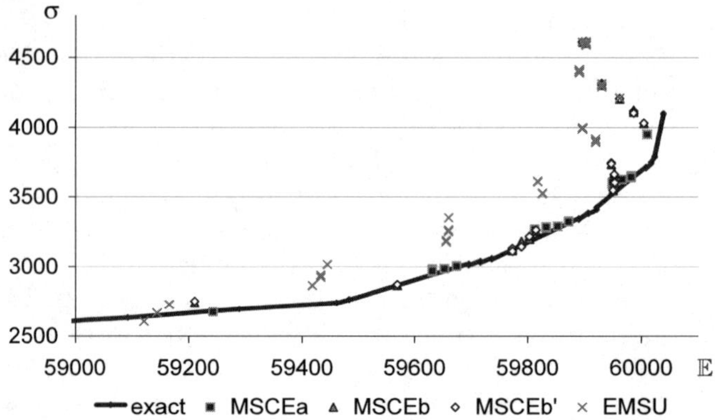

Fig. 7.15. Expected total revenue and standard deviation of total revenue given a γ-optimal policy and policies obtained by MSCE-a, MSCE-b, MSCE-b', and EMSU.

8

An Extension: Capacity Control Under a General Discrete Choice Model of Consumer Behavior

In this chapter, we use an example to show that structural properties known from the risk-neutral setting carry over to the setting of a decision-maker with a concave exponential atemporal utility function even for more general capacity control models.

In particular, we analyze an extension of the dynamic revenue management model under a general discrete choice model of consumer behavior by Talluri and van Ryzin (2004a) that was initially stated in terms of expected revenue maximization.

After a short revision of the model, we summarize the main results from the perspective of a risk-neutral decision-maker before we turn to the perspective of an expected utility maximizing decision-maker. We concentrate on the maximization of expected atemporal exponential utility.[1]

8.1 The Capacity Control Model

As in the dynamic model, time is discrete and time periods are indexed by n. Smaller values of n represent later points in time, i.e. the indices run backwards, and n equals the number of remaining periods. Accordingly, period $n = 0$ represents the end of the booking horizon; the beginning corresponds to index N. In each period, a maximum of one customer request arrives. There are $i_{\max}$ products, with $\Im = \{1, \ldots, i_{\max}\}$ denoting the entire set of products. Every product $i \in \Im$ has an associated fare ϱ_i, and without loss of generality, the products are indexed so that $\varrho_1 \geq \varrho_2 \geq \cdots \geq \varrho_{i_{\max}} \geq 0$.

At the beginning of each period n, the firm must choose a subset $\mathfrak{M} \subseteq \Im$ of products to offer. When the set of products $\mathfrak{M}$ is offered, $\hat{p}_i(\mathfrak{M})$ denotes the

[1] Parallel to this research, but independently, Feng and Gallego (2005) analyze expected utility maximizing capacity control policies for this model in a continuous time setting. Using a different line of argumentation, they obtain the same results as presented here.

probability that a customer arrives and chooses product $i \in \mathfrak{M}$. We assume $\hat{p}_i(\mathfrak{M}) = 0$ if $i \notin \mathfrak{M}$. The artificial class $i = 0$ represents the no-purchase choice, i.e. the event that no customer arrives or the customer does not purchase any of the fares offered in $\mathfrak{M}$ within the period considered. The probability $\hat{p}_0(\mathfrak{M}) = 1 - \sum_{i \in \mathfrak{M}} \hat{p}_i(\mathfrak{M})$ is the corresponding no-purchase probability. It is possible to allow the choice probabilities to be a function of the remaining time periods n, but to keep the notation simple, we assume time-independent probabilities.

In the dynamic model as introduced in Sect. 5.1, demand for a product is independent of the availability of other products cf. assumption x). Thus, the dynamic model arises as a special case, with $\hat{p}_i(\mathfrak{M}) = \hat{p}_i$ if $i \in \mathfrak{M}$ and 0 else. In this case, we can think about the problem in terms of simple accept or deny decisions.

Here, the only condition we impose on the choice probabilities $\hat{p}_i(\mathfrak{M})$ is that they define a proper probability function. That is, for every set $\mathfrak{M} \subseteq \mathfrak{I}$, the probabilities satisfy $\hat{p}_i(\mathfrak{M}) \geq 0$ for all $i \in \mathfrak{M}$ and $\sum_{i \in \mathfrak{M}} \hat{p}_i(\mathfrak{M}) + \hat{p}_0(\mathfrak{M}) = 1$. This includes most choice models of interest, among others the multinomial logit model; see Talluri and van Ryzin (2004b).

Again, we let C denote the total capacity and c the number of remaining units in a given period.

8.2 Maximizing Expected Revenue

Talluri and van Ryzin (2004a) reduce the objective of finding a policy that maximizes the expected revenue to solving the optimality equation of a finite-stage Markov decision model $(N, \mathfrak{X}, \mathfrak{A}_n, p_n, r_n^e, V_0)$ with planning horizon N. The state space is $\mathfrak{X} = \{c \in \mathbb{Z} \mid c \leq C\}$ and represents the remaining capacity. The action space is the power set of $\mathfrak{I}$, i.e. $\mathfrak{A} = \mathfrak{A}_n = \mathcal{P}(\mathfrak{I})$ for all n. Action a denotes the set of products offered, with $\mathfrak{A}(c) = \mathfrak{A}$ for all $c \geq 0$, $\mathfrak{A}(c) = \emptyset$ for $c < 0$. The transition laws p_n from $\mathfrak{K}_n = \{(c, a) \mid c \in \mathfrak{X}, a \in \mathfrak{A}\}$ into $\mathfrak{X}$ are defined by $p_n(c, a, c-1) = \sum_{i \in \mathfrak{M}} \hat{p}_i(\mathfrak{M})$, $p_n(c, a, c) = 1 - p_n(c, a, c-1)$, and 0 otherwise (with $\hat{p}_0(\mathfrak{M})$ arbitrary). The one-stage reward functions r_n on $\mathfrak{K}_n$ are bounded by ϱ_1, and have expected values of $r_n^e(c, a) = \sum_{i \in \mathfrak{M}} \hat{p}_i(\mathfrak{M}) \varrho_i$. The terminal reward function V_0 on $\mathfrak{X}$ can again be chosen as $V_0(c) = 0$ for $c \geq 0$ and $V_0(c) = \bar{\varrho} c$ for $c < 0$ with $\bar{\varrho} > \varrho_1$.

For each period n, the decision-maker must decide which set of products a_n to offer given the residual capacity c_n.

Let $(X_N, X_{N-1}, \ldots, X_0)$ denote the state process of the MDP and $\mathfrak{F}^N$ the set of all policies. The maximum expected revenue starting N periods before departure with capacity c,

$$V^*(c) = \max_{\pi \in \mathfrak{F}^N} \mathbb{E}_\pi \left[\sum_{n=1}^{N} r_n^e(X_n, f_n(X_n)) + V_0(X_0) \mid X_N = c \right],$$

can be obtained by backward induction on n for all $c \in \mathfrak{X}$ from the optimality equation

$$V_n(c) = \max_{\mathfrak{M} \in \mathfrak{A}(c)} \left\{ \sum_{i \in \mathfrak{M}} \hat{p}_i(\mathfrak{M}) \left[\varrho_i - \Delta V_{n-1}(c) \right] \right\} + V_{n-1}(c) \tag{8.1}$$

starting with V_0.

An easy induction on n using $\varrho_1 < \bar{\varrho}$ shows that $V_n(0) = 0$ and that $\mathfrak{M} = \emptyset$ in the case of $c \leq 0$. The proof is basically the same as the one stated in Lemma 5.2 (ii). We can thus enlarge the set of feasible actions to $\mathfrak{A}(c) = \mathfrak{A} = \mathcal{P}(\mathfrak{I})$ for all c.

If we let

$$\lambda(\mathfrak{M}) = \sum_{i \in \mathfrak{M}} \hat{p}_i(\mathfrak{M}) = 1 - \hat{p}_0(\mathfrak{M})$$

denote the total probability of purchase and

$$\omega(\mathfrak{M}) = \sum_{i \in \mathfrak{M}} \hat{p}_i(\mathfrak{M}) \varrho_i$$

the expected one-stage revenue from offering set $\mathfrak{M}$, (8.1) can be written in a more compact form as

$$V_n(c) = \max_{\mathfrak{M} \subseteq \mathfrak{I}} \left\{ \omega(\mathfrak{M}) - \lambda(\mathfrak{M}) \Delta V_{n-1}(c) \right\} + V_{n-1}(c) \ . \tag{8.2}$$

The sequence of sets achieving the maximum in (8.2) forms an optimal Markovian policy. Because the action space, i.e. the number of subsets, is finite, there is always one set $\mathfrak{M}$ that maximizes $\omega(\mathfrak{M}) - \lambda(\mathfrak{M}) \Delta V_{n-1}(c)$, so randomizing among the sets provides no additional benefit to the seller. Yet Talluri and van Ryzin (2004a) allow this flexibility in policies, since it is advantageous in the following analysis.

Basically, Talluri and van Ryzin (2004a) show that an ordered family of "efficient" subsets $\mathfrak{M}_1, \ldots, \mathfrak{M}_\mu$ can be identified. An optimal policy of this decision problem opens one of these efficient sets in each period. The more capacity available (or the less time), the further the optimal set is along this sequence. We revise their argumentation briefly.

8.2.1 Efficient Sets

Talluri and van Ryzin (2004a) define efficient sets as follows.

Definition 8.1. *A set $\mathfrak{N}$ is inefficient if there exist probabilities $\alpha(\mathfrak{M})$ for all $\mathfrak{M} \subseteq \mathfrak{I}$ (including the empty set $\mathfrak{M} = \emptyset$) with $\sum_{\mathfrak{M} \subseteq \mathfrak{I}} \alpha(\mathfrak{M}) = 1$ such that*

$$\lambda(\mathfrak{N}) \geq \sum_{\mathfrak{M} \subseteq \mathfrak{I}} \alpha(\mathfrak{M}) \lambda(\mathfrak{M}) \quad \textit{and} \quad \omega(\mathfrak{N}) < \sum_{\mathfrak{M} \subseteq \mathfrak{I}} \alpha(\mathfrak{M}) \omega(\mathfrak{M}) \ .$$

Otherwise, the set $\mathfrak{N}$ is efficient.

In addition, they introduce the notion of an efficient frontier, which is frequently used in parametric linear programming. The following lemma is a standard result; see e.g. Theorem 5.1 in Bertsimas and Tsitsiklis (1997, p. 213).

Lemma 8.1. *The efficient frontier $F : [0,1] \to \mathbb{R}$, defined by*

$$F(b) = \max\left\{ \sum_{\mathfrak{M}\subseteq\mathfrak{I}} \alpha(\mathfrak{M})\omega(\mathfrak{M}) : \sum_{\mathfrak{M}\subseteq\mathfrak{I}} \alpha(\mathfrak{M})\lambda(\mathfrak{M}) \leq b, \right.$$

$$\left. \sum_{\mathfrak{M}\subseteq\mathfrak{I}} \alpha(\mathfrak{M}) = 1, \alpha(\mathfrak{M}) \geq 0 \text{ for all } \mathfrak{M} \subseteq \mathfrak{I} \right\},$$

is increasing and concave in b.

In addition, Talluri and van Ryzin (2004a) prove the subsequent results.

Lemma 8.2. *A set $\mathfrak{N}$ is efficient if and only if for some value $\xi \geq 0$, $\mathfrak{N}$ is an optimal solution to*

$$\max_{\mathfrak{M}\subseteq\mathfrak{I}} \{\omega(\mathfrak{M}) - \xi\lambda(\mathfrak{M})\}\ .$$

Proposition 8.1. *An inefficient set is never an optimal solution to (8.2).*

8.2.2 Structure of an Optimal Policy

Let μ denote the number of efficient sets. From now on, we assume the efficient sets, $\mathfrak{M}_1, \ldots, \mathfrak{M}_\mu$, are indexed in increasing order of expected one-stage utility and total probability of purchase. This indexing is possible because the efficient frontier is increasing. Let $\omega_j = \omega(\mathfrak{M}_j)$ and $\lambda_j = \lambda(\mathfrak{M}_j)$. Using this notational simplification, the Bellman equation reads

$$V_n(c) = \max_{j=1,\ldots,\mu} \{\omega_j - \lambda_j \Delta V_{n-1}(c)\} + V_{n-1}(c)\ . \tag{8.3}$$

Talluri and van Ryzin (2004a) show the following relationship between the optimal efficient set and $\Delta V_{n-1}(c)$.

Lemma 8.3. *The index of the efficient set that maximizes (8.3) (or greatest such index if more than one efficient set maximizes (8.3)) is decreasing in $\Delta V_{n-1}(c)$.*

In addition, they show that $\Delta V_n(c)$ is monotone in n and c.

Lemma 8.4. *For all $c \in \mathfrak{C}$ and $n = 1, \ldots, N$,*

(i) $\Delta V_n(c) \leq \Delta V_n(c-1)$,
(ii) $\Delta V_{n-1}(c) \leq \Delta V_n(c)$.

These lemmas can be used to show the following theorem:

Theorem 8.1. *An optimal policy for (8.1) can be found by selecting a set* $\mathfrak{M}_{j^*}$ *from among the* μ *efficient, ordered sets* $\{\mathfrak{M}_j, j = 1, \ldots \mu\}$ *that maximizes (8.3). Moreover, for fixed* n*, the largest optimal index* j^* *is increasing in the remaining capacity* c*, and for any fixed* c*,* j^* *is decreasing in the remaining periods* n*.*

Consequently, the search for an optimal set can be reduced to efficient sets, which in many cases significantly decreases the number of sets that need to be considered. This limited number of sets can be sequenced in a natural way so that the more capacity one has (or the fewer periods remaining), the higher the index of the set one should use.

In general, the task of identifying the efficient sets is still computationally complex. The naive approach is to enumerate all $2^{\imath_{\max}} - 1$ subsets of $\Im$ and test each for efficiency. Talluri and van Ryzin (2004a) propose a more efficient alternative, which they call the largest marginal revenue procedure. This idea exploits the fact that the efficient frontier is known to be concave.

First, let $\mathfrak{M}_0 = \emptyset$. Given the first j efficient sets, the $j+1$st efficient set can be found by checking among the sets $\mathfrak{M}$ with $\lambda(\mathfrak{M}) \geq \lambda(\mathfrak{M}_j)$ and $\omega(\mathfrak{M}) \geq \omega(\mathfrak{M}_j)$ for the one that maximizes the marginal revenue ratio

$$\frac{\omega(\mathfrak{M}) - \omega(\mathfrak{M}_j)}{\lambda(\mathfrak{M}) - \lambda(\mathfrak{M}_j)} .$$

The procedure starts with $j = 0$ and stops when no sets with $\lambda(\mathfrak{M}) \geq \lambda(\mathfrak{M}_j)$ and $\omega(\mathfrak{M}) \geq \omega(\mathfrak{M}_j)$ exist. Since there are $O(2^{\imath_{\max}})$ subsets to check at each step, the recursion has a complexity of $O(\mu 2^{\imath_{\max}})$. For other heuristic and analytic methods to reduce the complexity, see Talluri and van Ryzin (2004a).

8.2.3 A Numerical Example

The following example is adapted from Talluri and van Ryzin (2004a). We consider an airline that offers three products, 1, 2, and 3 with associated revenues of 800, 500, and 450, respectively. These products differ not only in revenue but also in conditions and restrictions imposed. Due to these restrictions, customer groups are segmented, but not perfectly. (This is where the model differs from the dynamic model in Sect. 5.1.) The resulting choice probabilities from offering set $\mathfrak{M}$ are given in Table 8.1.

In the scatter plot, Fig. 8.1, one can clearly see that the only efficient sets are $\mathfrak{M}_0 = \emptyset \subset \mathfrak{M}_1 = \{1\} \subset \mathfrak{M}_2 = \{1,3\} \subset \mathfrak{M}_3 = \{1,2,3\}$. These sets are marked with an asterisk in Table 8.1. In general, a policy with $\mathfrak{M}_0 \subseteq \mathfrak{M}_1 \subseteq \cdots \subseteq \mathfrak{M}_\mu$ is called a nested policy. Given such a nested optimal policy, the optimal protection levels for higher indexed sets $j = 1, \ldots, \mu$ are

$$y^*_{j-1}(n) = \max\{c \in \mathbb{N}_0 \ : \ \omega_{j-1} - \lambda_{j-1}\Delta V_{n-1}(c) > \omega_j - \lambda_j \Delta V_{n-1}(c)\} .$$

Table 8.1. Choice probabilities $\hat{p}_i(\mathfrak{M})$, total probability of purchase $\lambda(\mathfrak{M})$, and expected one-stage revenues $\omega(\mathfrak{M})$.

$\mathfrak{M}$	$\hat{p}_1(\mathfrak{M})$	$\hat{p}_2(\mathfrak{M})$	$\hat{p}_3(\mathfrak{M})$	$\hat{p}_0(\mathfrak{M})$	$\lambda(\mathfrak{M})$	$\omega(\mathfrak{M})$
$\{\emptyset\}$	0	0	0	1.0	0	0*
$\{1\}$	0.3	0	0	0.7	0.3	240*
$\{2\}$	0	0.4	0	0.6	0.4	200
$\{3\}$	0	0	0.5	0.5	0.5	225
$\{1,2\}$	0.1	0.6	0	0.3	0.7	380
$\{1,3\}$	0.3	0	0.5	0.2	0.8	465*
$\{2,3\}$	0	0.4	0.5	0.1	0.9	425
$\{1,2,3\}$	0.1	0.4	0.5	0	1.0	505*

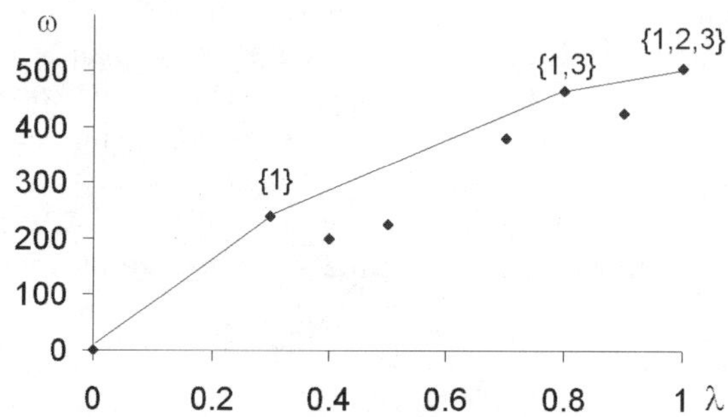

Fig. 8.1. Scatter plot of λ and ω as well as the efficient sets.

But note that in contrast to the basic dynamic model, the nesting need not be by fare class, as can be seen in this example.

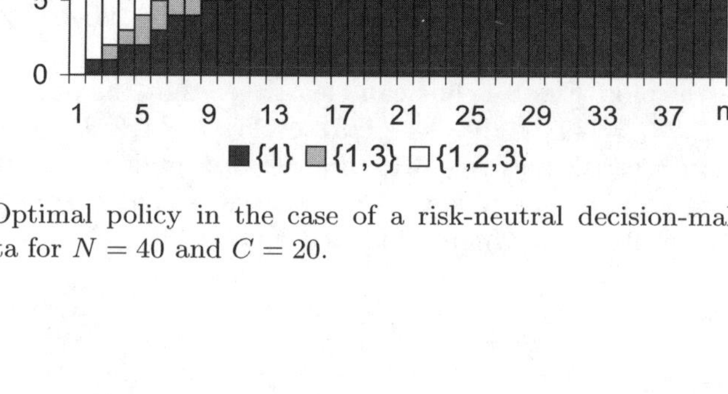

Fig. 8.2. Optimal policy in the case of a risk-neutral decision-maker given the example data for $N = 40$ and $C = 20$.

Figure 8.2 illustrates an optimal policy for $N = 40$ and $C = 20$. In particular, the protection levels of the optimal policy can be read from this chart. Consider e.g. the situation $n = 11$ periods before departure. If the remaining capacity is 11 or more, all products, set $\{1, 2, 3\}$, should be offered. If the capacity is 6 or less, the set $\{1\}$, consisting of only product 1, is the optimal action. For 7 to 10 remaining seats, the set $\{1, 3\}$ should be offered.

In line with the results mentioned above, the optimal actions comprise efficient sets only. In addition, the index of the set offered is monotone in the remaining number of periods and the remaining capacity.

8.3 Maximizing Expected Utility

As in the chapters before, we revise the capacity control model under a general discrete choice model of consumer behavior from the perspective of a risk-averse decision-maker.

Note that an additive time-separable utility function results in a pure rescaling of the product fares. Since the time periods of the dynamic model are assumed to be small, however, we concentrate on atemporal utility functions in the following.

In Sect. 3.2.2, we said that using MDPs, the standard approach for maximizing expected utility with an atemporal utility function u_0 is to introduce an additional wealth variable w and use the optimality equation given in (3.11).

When the set $\mathfrak{M}$ is offered, one unit of capacity is sold with probability $\hat{p}_i(\mathfrak{M})$ at a price ϱ_i. Hence, the probability of a transition from state (c, w) to $(c - 1, w + \varrho_i)$ when set $\mathfrak{M}$ is offered is $\hat{p}_i(\mathfrak{M})$, and the probability for a transition to (c, w) is $1 - \sum_{i \in \mathfrak{M}} \hat{p}_i(\mathfrak{M})$; all other transition probabilities are 0. Using these transition probabilities $p_n(c, w, a, c', w')$ in the optimality equation (3.11), we obtain that

$$V^{\text{atmp}*}(c, w) = \max_{\pi \in \mathfrak{F}^N} \mathbb{E}_\pi \left[u_0(W_0 + V_0(X_0)) \mid X_N = c, W_N = w \right]$$

is the unique solution of

$$\begin{aligned} V_n^{\text{atmp}}(c, w) = & \max_{\mathfrak{M} \in \mathfrak{A}(c)} \left\{ \sum_{i \in \mathfrak{M}} \hat{p}_i(\mathfrak{M}) \left[V_{n-1}^{\text{atmp}}(c - 1, w + \varrho_i) - V_{n-1}^{\text{atmp}}(c, w) \right] \right\} \\ & + V_{n-1}^{\text{atmp}}(c, w) , \end{aligned} \tag{8.4}$$

with $V_0^{\text{atmp}}(c, w) = u_0(V_0(c) + w)$. Every policy $\pi^{\text{atmp}*}$ formed by actions maximizing the right-hand side of (8.4), is atmp-optimal, i.e. leads to $V^{\text{atmp}*}$.

As in the basic dynamic model of Sect. 5.1, monotonicity of the value function in c and w follows by induction, and the same counterexamples for monotonicity of an optimal policy in c and w given general increasing utility functions can be used. For an exponential utility function

$$u_\gamma(w) = -\exp(-\gamma w), \qquad w \in \mathbb{R} ,$$

with $\gamma > 0$, however, a γ-optimal policy for the dynamic model has the same structure as an expected revenue maximizing policy in the model of Sect. 5.1. Now we examine if this is also true for the capacity control model under a general discrete choice model of consumer behavior.

To simplify (8.4), the delta-property of the exponential utility function and the same line of argumentation as in Sect. 3.2.3 can be used. This yields that the policy that achieves the maximum expected utility given c seats remaining n periods before departure can be obtained by backward induction from

$$V_n^\gamma(c) = \max_{\mathfrak{M}\in\mathfrak{A}(c)} \left\{ \sum_{i\in\mathfrak{M}} \hat{p}_i(\mathfrak{M}) \left[\exp(-\gamma\varrho_i) V_{n-1}^\gamma(c-1) - V_{n-1}^\gamma(c)\right] \right\} + V_{n-1}^\gamma(c) , \tag{8.5}$$

with initial value $V_0^\gamma(c) = u_\gamma(V_0(c))$. As in the risk-neutral setting, it follows by induction on n that $V_n^\gamma(0) = -1$ and that in the case of $c \leq 0$, the γ-optimal action to offer is the empty set $\emptyset$.

Lemma 8.5. *For $c \leq 0$, $V_n^\gamma(c) = -\exp(-\gamma\bar{\varrho}c)$ for all $n = 0, \ldots, N$.*

Proof. By definition, $V_0^\gamma(c) = -\exp(-\gamma\bar{\varrho}c)$, so the assertion is true for $n = 0$. Assume it is true for some n; then

$$V_{n+1}^\gamma(c) = \max_{\mathfrak{M}\subseteq\mathfrak{I}} \left\{ -\exp(-\gamma\bar{\varrho}c) \sum_{i\in\mathfrak{M}} \hat{p}_i(\mathfrak{M}) \left[\exp(\gamma(\bar{\varrho}-\varrho_i)) - 1\right] \right\} - \exp(-\gamma\bar{\varrho}c) .$$

Since $\bar{\varrho} > \varrho_1$, the maximum is attained when the action $\mathfrak{M} = \emptyset$ is chosen with $V_{n+1}^\gamma(c) = -\exp(-\gamma\bar{\varrho}c)$. □

Consequently, we can enlarge the set of feasible actions in state c again to $\mathfrak{A}(c) = \mathfrak{A} = \mathcal{P}(\mathfrak{I})$ for all c. As before, we let

$$\lambda(\mathfrak{M}) = \sum_{i\in\mathfrak{M}} \hat{p}_i(\mathfrak{M}) = 1 - \hat{p}_0(\mathfrak{M})$$

denote the total probability of purchase and

$$\omega^\gamma(\mathfrak{M}) = \sum_{i\in\mathfrak{M}} \hat{p}_i(\mathfrak{M}) u_\gamma(\varrho_i) + \hat{p}_0(\mathfrak{M}) u_\gamma(0)$$

the expected one-stage utility from offering set $\mathfrak{M}$. (8.5) can then be rearranged to

$$V_n^\gamma(c) = \Delta V_{n-1}^\gamma(c) - V_{n-1}^\gamma(c-1) \cdot \max_{\mathfrak{M}\subseteq\mathfrak{I}} \left\{ \omega^\gamma(\mathfrak{M}) - \lambda(\mathfrak{M}) \left(1 - \Gamma V_{n-1}^\gamma(c)\right) \right\} , \tag{8.6}$$

as $-V_{n-1}^{\gamma}(c-1) > 0$. An argument similar to the one used in the discussion on the MSCE heuristics shows that the term $-1/\gamma \ln \Gamma V_{n-1}^{\gamma}(c)$ can be interpreted as the marginal certainty equivalent of the cth seat in period n. Then, $-\Gamma V_{n-1}^{\gamma}(c)$ is the utility of this marginal certainty equivalent.

A sequence of sets achieving the maximum in (8.6) forms a γ-optimal Markovian policy. Randomizing among the sets provides no additional benefit to the seller, but is again theoretically useful.

8.3.1 Efficient Sets

Efficient sets and the efficient frontier can be defined as before, with ω^{γ} written in place of ω. Given a fixed γ, the γ-efficient frontier $F^{\gamma}: [0,1] \to \mathbb{R}$ is increasing and concave. Furthermore, a set $\mathfrak{N}$ is γ-efficient if and only if for some value $\xi \geq 0$, the set $\mathfrak{N}$ is a γ-optimal solution to

$$\max_{\mathfrak{M} \subseteq \mathfrak{F}} \{\omega^{\gamma}(\mathfrak{M}) - \xi \lambda(\mathfrak{M})\} .$$

Since, by definition of the exponential utility function, $1 - \Gamma V_{n-1}^{\gamma}(c)$ is nonnegative, it follows that a γ-inefficient set is never a γ-optimal solution to (8.6).

8.3.2 Characterization of the Optimal Policy

We assume again that for a fixed γ, the γ-efficient sets are denoted by $\mathfrak{M}_1, \ldots, \mathfrak{M}_{\mu}$ and are indexed in increasing order of expected one-stage utility and total probability of purchase. Note, however, that in general the number of γ-efficient sets, μ, generally depends on γ.

As before, let $\omega_j^{\gamma} = \omega^{\gamma}(\mathfrak{M}_j)$ and $\lambda_j = \lambda(\mathfrak{M}_j)$. This yields

$$V_n^{\gamma}(c) = \Delta V_{n-1}^{\gamma}(c) - V_{n-1}^{\gamma}(c-1) \max_{j=1,\ldots,\mu} \left\{\omega_j^{\gamma} - \lambda_j \left(1 - \Gamma V_{n-1}^{\gamma}(c)\right)\right\} . \quad (8.7)$$

In order to show that the γ-efficient sets are used in the order of increasing selling probability, we need the following properties of $-\Gamma V_n^{\gamma}(c)$. The first one is that the utility of the marginal certainty equivalent of the cth seat is decreasing in the remaining capacity c, which is intuitive. The second property is that $-\Gamma V_n^{\gamma}(c)$ is monotone in the number of remaining periods as well.

Lemma 8.6. $\Gamma V_n^{\gamma}(c) \geq \Gamma V_n^{\gamma}(c-1), \quad n = 1, \ldots, N, \; c = 1, \ldots, C.$

Proof. The proof is by induction on n. First, the statement is trivially true for $n = 0$ by the definition of the terminal reward. Assume it is true for period $n-1$, i.e. $V_{n-1}^{\gamma}(c)V_{n-1}^{\gamma}(c-2) - V_{n-1}^{\gamma}(c-1)^2 \geq 0$ for all $c = 2, 3 \ldots, C$. Let $\mathfrak{M}_n^*(c-1)$ denote the optimal solution to (8.6) in period n given a capacity of $c-1$ and write $\hat{\omega} = \omega^{\gamma}(\mathfrak{M}_n^*(c-1))$ and $\hat{\lambda} = \lambda(\mathfrak{M}_n^*(c-1))$ for short.

Now, using $\hat{\omega}$ and $\hat{\lambda}$, one can rearrange (8.6) and obtain

$$
\begin{aligned}
&V_n^\gamma(c)V_n^\gamma(c-2) - V_n^\gamma(c-1)^2 \\
&\geq \left[(1-\hat{\lambda})V_{n-1}^\gamma(c) - (1+\hat{\omega}-\hat{\lambda})V_{n-1}^\gamma(c-1)\right] \\
&\quad \cdot \left[(1-\hat{\lambda})V_{n-1}^\gamma(c-2) - (1+\hat{\omega}-\hat{\lambda})V_{n-1}^\gamma(c-3)\right] \\
&\quad - \left[(1-\hat{\lambda})V_{n-1}^\gamma(c-1) - (1+\hat{\omega}-\hat{\lambda})V_{n-1}^\gamma(c-2)\right]^2 ,
\end{aligned}
$$

where the inequality follows from the fact that $\hat{\lambda}$ and $\hat{\omega}$ are maximizing for a remaining capacity of $c-1$ and not necessarily for c and $c-2$.

Rearranging and canceling terms yields

$$
\begin{aligned}
&V_n^\gamma(c)V_n^\gamma(c-2) - V_n^\gamma(c-1)^2 \\
&\geq (1-\hat{\lambda})(\hat{\lambda}-1-\hat{\omega})\left[V_{n-1}^\gamma(c)V_{n-1}^\gamma(c-3) - V_{n-1}^\gamma(c-2)V_{n-1}^\gamma(c-1)\right] \\
&\quad + (1+\hat{\omega}-\hat{\lambda})^2\left[V_{n-1}^\gamma(c-1)V_{n-1}^\gamma(c-3) - V_{n-1}^\gamma(c-2)^2\right] \\
&\quad + (1-\hat{\lambda})^2\left[V_{n-1}^\gamma(c)V_{n-1}^\gamma(c-2) - V_{n-1}^\gamma(c-1)^2\right] .
\end{aligned}
$$

Since total probabilities of purchase are smaller than 1, it holds that $1 - \lambda(\mathfrak{M}) = \hat{p}_0(\mathfrak{M}) \geq 0$ for all $\mathfrak{M} \subseteq \mathfrak{I}$. Furthermore,

$$
\lambda(\mathfrak{M}) - 1 - \omega^\gamma(\mathfrak{M}) = -\sum_{i\in\mathfrak{M}} \hat{p}_i(\mathfrak{M})u_\gamma(\varrho_i) = \sum_{i\in\mathfrak{M}} \hat{p}_i(\mathfrak{M})\exp(-\gamma\varrho_i) \geq 0
$$

for all $\mathfrak{M} \subseteq \mathfrak{I}$. All other factors are non-negative due to the induction hypothesis. Hence,

$$
V_n^\gamma(c)V_n^\gamma(c-2) - V_n^\gamma(c-1)^2 \geq 0 ,
$$

which is equivalent to $\Gamma V_n^\gamma(c) \geq \Gamma V_n^\gamma(c-1)$. □

Lemma 8.7. $\Gamma V_{n-1}^\gamma(c) \leq \Gamma V_n^\gamma(c), \quad n = 2,\ldots,N, c = 1,\ldots,C.$

Proof. Rearranging the hypothesis yields the equivalent statement

$$
\frac{V_n^\gamma(c)}{V_{n-1}^\gamma(c)} \leq \frac{V_n^\gamma(c-1)}{V_{n-1}^\gamma(c-1)} .
$$

Substituting (8.5) for $V_n^\gamma(c)$ gives

$$
\begin{aligned}
\frac{V_n^\gamma(c)}{V_{n-1}^\gamma(c)} &= \max_{\mathfrak{M}\subseteq\mathfrak{I}}\left\{\sum_{i\in\mathfrak{M}} \hat{p}_i(\mathfrak{M})\left[\frac{\exp(-\gamma\varrho_i)}{\Gamma V_{n-1}^\gamma(c)} - 1\right]\right\} + 1 \\
&\leq \max_{\mathfrak{M}\subseteq\mathfrak{I}}\left\{\sum_{i\in\mathfrak{M}} \hat{p}_i(\mathfrak{M})\left[\frac{\exp(-\gamma\varrho_i)}{\Gamma V_{n-1}^\gamma(c-1)} - 1\right]\right\} + 1 \\
&= \frac{V_n^\gamma(c-1)}{V_{n-1}^\gamma(c-1)}
\end{aligned}
$$

for the right-hand side, with the inequality following from Lemma 8.6. □

By combining Lemma 8.3 with Lemmas 8.6 and 8.7, we obtain the following theorem.

Theorem 8.2. *Given a fixed γ, a γ-optimal policy for (8.5) can be found by selecting a set $\mathfrak{M}_{j^*}$ from among the γ-efficient, ordered sets $\{\mathfrak{M}_j : j = 1, \ldots, \mu\}$ that maximizes (8.7). Moreover, for fixed n, the largest optimal index j^* is increasing in the remaining capacity c, and for any fixed c, j^* is decreasing in the number of remaining periods n.*

In order to spot these γ-efficient sets, one could substitute expected revenues for expected one-stage utilities in the "largest marginal revenue" procedure to yield a "largest marginal expected utility" procedure using the ratio

$$\frac{\omega^\gamma(\mathfrak{M}) - \omega^\gamma(\mathfrak{M}_j)}{\lambda(\mathfrak{M}) - \lambda(\mathfrak{M}_j)} \,.$$

instead of the marginal revenue ratio.

We can show the following relationship between the risk-sensitive and the risk-neutral case.

Proposition 8.2. *A policy that is γ-optimal for all $\gamma \in (0, \gamma_0)$ is an expected revenue maximizing policy.*

Proof. First, note that from the optimality equation (8.5), we can directly see that an efficient set $\mathfrak{M}^*$ is an optimal action (there might be more than one) if

$$\begin{aligned}&\sum_{i\in\mathfrak{M}^*} \hat{p}_i(\mathfrak{M}^*)\left[\exp(-\gamma\varrho_i)V_{n-1}^\gamma(c-1) - V_{n-1}^\gamma(c)\right]\\ &\geq \sum_{i\in\mathfrak{M}} \hat{p}_i(\mathfrak{M})\left[\exp(-\gamma\varrho_i)V_{n-1}^\gamma(c-1) - V_{n-1}^\gamma(c)\right]\end{aligned}$$

for all $\mathfrak{M} \in \mathcal{P}(\Im)$.

Solving for $\Gamma V_n^\gamma(c)$, taking logarithms, and multiplying by $(-1/\gamma)$ yields that an efficient set $\mathfrak{M}^*$ is an optimal action if for all sets $\mathfrak{M} \subseteq \Im$ with $\lambda(\mathfrak{M}^*) - \lambda(\mathfrak{M}) > 0$,

$$\begin{aligned}&-\frac{1}{\gamma}\ln\left(\frac{\sum_{i\in\mathfrak{M}^*}\hat{p}_i(\mathfrak{M}^*)\exp(-\gamma\varrho_i) - \sum_{i\in\mathfrak{M}}\hat{p}_i(\mathfrak{M})\exp(-\gamma\varrho_i)}{\sum_{i\in\mathfrak{M}^*}\hat{p}_i(\mathfrak{M}^*) - \sum_{i\in\mathfrak{M}}\hat{p}_i(\mathfrak{M})}\right)\\ &\geq -\frac{1}{\gamma}\ln\left(\Gamma V_n^\gamma(c)\right) ,\end{aligned} \tag{8.8}$$

and for all sets $\mathfrak{M} \subseteq \Im$ with $\lambda(\mathfrak{M}^*) - \lambda(\mathfrak{M}) < 0$,

$$\begin{aligned}&-\frac{1}{\gamma}\ln\left(\frac{\sum_{i\in\mathfrak{M}^*}\hat{p}_i(\mathfrak{M}^*)\exp(-\gamma\varrho_i) - \sum_{i\in\mathfrak{M}}\hat{p}_i(\mathfrak{M})\exp(-\gamma\varrho_i)}{\sum_{i\in\mathfrak{M}^*}\hat{p}_i(\mathfrak{M}^*) - \sum_{i\in\mathfrak{M}}\hat{p}_i(\mathfrak{M})}\right)\\ &\leq -\frac{1}{\gamma}\ln\left(\Gamma V_n^\gamma(c)\right) .\end{aligned} \tag{8.9}$$

If there is an efficient set $\mathfrak{M}$ with $\lambda(\mathfrak{M}^*) = \lambda(\mathfrak{M})$, their expected utilities must be equal, and both actions, $\mathfrak{M}$ and $\mathfrak{M}^*$, are optimal.

Denote by $\mathfrak{F}_\emptyset$ the finite set of all decision rules with $f_n(c) = \emptyset$ for $c \leq 0$. As mentioned before, the expected revenue maximizing optimal policy is in $\mathfrak{F}_\emptyset^N$. The γ-optimal policy is in $\mathfrak{F}_\emptyset^N$ as well according to Lemma 8.5.

Now apply the Taylor approximation (3.6) to the expected utility from applying a policy $\pi \in \mathfrak{F}_0^N$. It follows that for $\gamma \to 0$,

$$
\begin{aligned}
-\frac{1}{\gamma}\ln(-V_n^\gamma(c)) &= \max_{\pi\in\mathfrak{F}_\emptyset^N} -\frac{1}{\gamma}\ln\left(\mathbb{E}_\pi\left[-\exp(-\gamma\,(W_0+V_0(X_0)))\mid X_N=c\right]\right)\\
&\to \max_{\pi\in\mathfrak{F}_\emptyset^N} \mathbb{E}_\pi\left[W_0+V_0(X_0)\mid X_N=c\right] = V_n(c)\,.
\end{aligned}
$$

In addition, note that another application of $\exp(x) = 1 + x + O(x^2)$ and $\ln(1+x) = x + O(x^2/2)$ for $x \to 0$ yields

$$
\begin{aligned}
&-\frac{1}{\gamma}\ln\left(\frac{\sum_{i\in\mathfrak{M}^*}\hat{p}_i(\mathfrak{M}^*)\exp(-\gamma\varrho_i)-\sum_{i\in\mathfrak{M}}\hat{p}_i(\mathfrak{M})\exp(-\gamma\varrho_i)}{\sum_{i\in\mathfrak{M}^*}\hat{p}_i(\mathfrak{M}^*)-\sum_{i\in\mathfrak{M}}\hat{p}_i(\mathfrak{M})}\right)\\
&=-\frac{1}{\gamma}\ln\left(\frac{\sum_{i\in\mathfrak{M}^*}\hat{p}_i(\mathfrak{M}^*)(1-\gamma\varrho_i+O(\gamma^2))}{\sum_{i\in\mathfrak{M}^*}\hat{p}_i(\mathfrak{M}^*)-\sum_{i\in\mathfrak{M}}\hat{p}_i(\mathfrak{M})}\right.\\
&\qquad\left.-\frac{\sum_{i\in\mathfrak{M}}\hat{p}_i(\mathfrak{M})(1-\gamma\varrho_i+O(\gamma^2))}{\sum_{i\in\mathfrak{M}^*}\hat{p}_i(\mathfrak{M}^*)-\sum_{i\in\mathfrak{M}}\hat{p}_i(\mathfrak{M})}\right)\\
&=-\frac{1}{\gamma}\ln\left(1-\gamma\frac{\sum_{i\in\mathfrak{M}^*}\hat{p}_i(\mathfrak{M}^*)\varrho_i-\sum_{i\in\mathfrak{M}}\hat{p}_i(\mathfrak{M})\varrho_i+O(\gamma^2)}{\sum_{i\in\mathfrak{M}^*}\hat{p}_i(\mathfrak{M}^*)-\sum_{i\in\mathfrak{M}}\hat{p}_i(\mathfrak{M})}\right)\\
&=\frac{\omega(\mathfrak{M}^*)-\omega(\mathfrak{M})}{\lambda(\mathfrak{M}^*)-\lambda(\mathfrak{M})}+O(\gamma)\,.
\end{aligned}
$$

Thus, for $\gamma \to 0$, the inequalities (8.8) and (8.9) converge to

$$\frac{\omega(\mathfrak{M}^*)-\omega(\mathfrak{M})}{\lambda(\mathfrak{M}^*)-\lambda(\mathfrak{M})} \geq \Delta V_n(c)$$

for all sets $\mathfrak{M} \subseteq \mathfrak{I}$ with $\lambda(\mathfrak{M}^*) - \lambda(\mathfrak{M}) > 0$ and to

$$\frac{\omega(\mathfrak{M}^*)-\omega(\mathfrak{M})}{\lambda(\mathfrak{M}^*)-\lambda(\mathfrak{M})} \leq \Delta V_n(c)$$

for all sets $\mathfrak{M} \subseteq \mathfrak{I}$ with $\lambda(\mathfrak{M}^*) - \lambda(\mathfrak{M}) < 0$. These inequalities determine optimal actions in the expected revenue maximizing case according to the (risk-neutral) optimality equation (8.2). □

8.3.3 A Numerical Example

For the example stated in the risk-neutral case above, the data as well as the values of $\omega^\gamma(\mathfrak{M})$ in the case of $\gamma = 0.001$ and 0.01 are summarized in Table

Table 8.2. Choice probabilities $\hat{p}_i(\mathfrak{M})$, total probability of purchase $\lambda(\mathfrak{M})$, and expected one-stage utilities $\omega^\gamma(\mathfrak{M})$ with $\gamma = 0.001$ and $\gamma = 0.01$.

$\mathfrak{M}$	$\hat{p}_1(\mathfrak{M})$	$\hat{p}_2(\mathfrak{M})$	$\hat{p}_3(\mathfrak{M})$	$\hat{p}_0(\mathfrak{M})$	$\lambda(\mathfrak{M})$	$\omega^{0.001}(\mathfrak{M})$	$\omega^{0.01}(\mathfrak{M})$
$\{\emptyset\}$	0	0	0	1.0	0	-1*	-1*
$\{1\}$	0.3	0	0	0.7	0.3	-0.8348*	-0.7001*
$\{2\}$	0	0.4	0	0.6	0.4	-0.8426	-0.6027
$\{3\}$	0	0	0.5	0.5	0.5	-0.8188	-0.5056
$\{1,2\}$	0.1	0.6	0	0.3	0.7	-0.7089	-0.3041*
$\{1,3\}$	0.3	0	0.5	0.2	0.8	-0.6536*	-0.2057
$\{2,3\}$	0	0.4	0.5	0.1	0.9	-0.6614	-0.1082
$\{1,2,3\}$	0.1	0.4	0.5	0	1.0	-0.6064*	-0.0083*

8.2. Figure 8.3 shows a scatter plot of $\lambda(\mathfrak{M})$, $\omega^{0.001}(\mathfrak{M})$, and $\omega^{0.01}(\mathfrak{M})$. In addition, the γ-efficient frontiers are indicated.

From the scatter plot, one can conclude that for $\gamma = 0.001$ the γ-efficient sets are the same as in the expected revenue maximizing case. For a more risk-averse decision-maker with $\gamma = 0.01$, however, the γ-efficient sets change to $\mathfrak{M}_0 = \emptyset \subset \mathfrak{M}_1 = \{1\} \subset \mathfrak{M}_2 = \{1,2\} \subset \mathfrak{M}_3 = \{1,2,3\}$. The γ-efficient sets for $\gamma = 0.01$ are difficult to spot in the scatter plot but can be obtained by minor calculations. In this setting, the sets are nested by fare class. The fact that for increasing absolute risk-aversion, i.e. values of γ, the set $\{1,2\}$ becomes more and more attractive over $\{1,3\}$ is plausible, since the variability in outcomes for $\{1,2\}$ is smaller than for $\{1,3\}$.

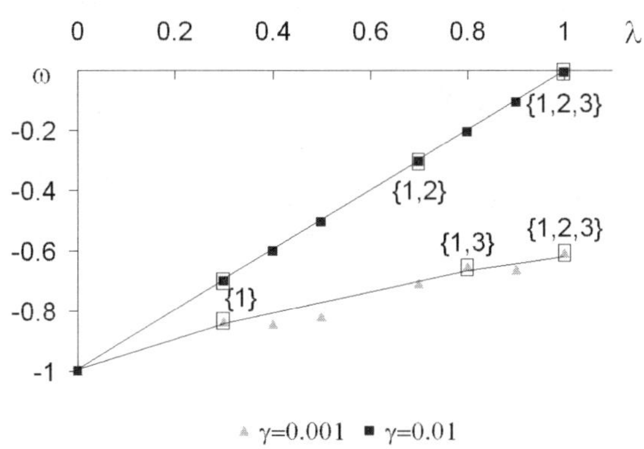

Fig. 8.3. Scatter plot of λ, $\omega^{0.001}$, and $\omega^{0.01}$ as well as the efficient frontiers.

For the case of $N = 40$ and $C = 20$, the γ-optimal actions, i.e. sets to offer, are depicted in Fig. 8.4 for $\gamma = 0.001$ and in Fig. 8.5 for $\gamma = 0.01$. As proven in Theorem 8.2, the γ-optimal actions comprise γ-efficient sets only. The index of the offered set is monotone in the remaining periods and remaining capacity. For a relatively low level of risk-aversion with the same γ-efficient sets as in the expected revenue maximizing case, the optimal and the γ-optimal policy are very similar. But note that in this example, the γ-optimal policy with $\gamma = 0.001$ tends to offer sets with a higher index.

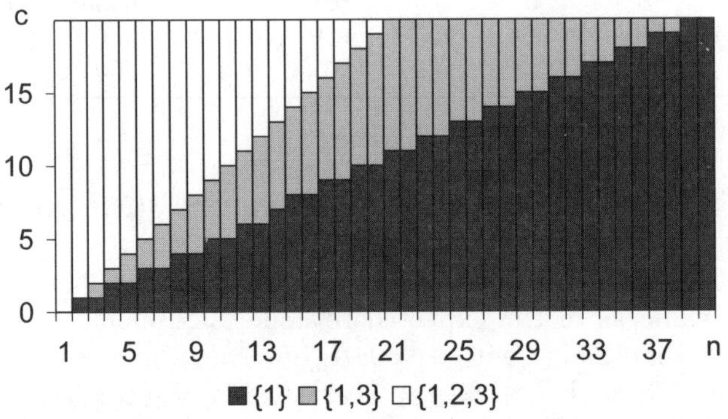

Fig. 8.4. γ-optimal policy with $\gamma = 0.001$.

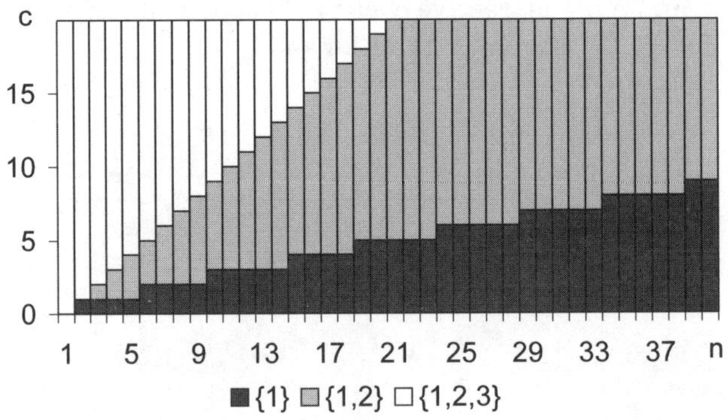

Fig. 8.5. γ-optimal policy with $\gamma = 0.01$.

9

Conclusion

We have presented (1) an expected revenue maximizing capacity control model that evolves in a random environment and (2) basic single-resource capacity control problems from the perspective of a risk-averse decision-maker.

The first model is an extension of existing capacity control and overbooking models. It is suitable for companies with many independent repetitions of the booking process if the occurrence of demand is known to depend on a random environment and if there is sufficient booking data to estimate the additional parameters.

The second approach is advisable for companies that apply capacity control but have only a few repetitions of the booking process, so that one single realization could potentially severely impact company revenues. For the two textbook models of capacity control, we have proven that in practice, a decision-maker with constant absolute risk-aversion (and no preferences towards the timing of revenue within the booking horizon) can adhere to the control by protection levels, which are well-known from the risk-neutral setting; their calculation simply has to be adapted.

9.1 Summary

After a review of Markov decision processes with the total reward criterion and expected utility theory for sequential decision making, we presented a capacity control model in a random environment with cancelations and no-shows. This model generalizes the omnibus model of Lautenbacher and Stidham (1999) as well as the previous combined capacity control and overbooking models of Subramanian et al. (1999) and Talluri and van Ryzin (2004b, pp. 155–161) by incorporating additional external factors that may have some impact on the distribution of the number and type of request arrivals. In contrast to the major part of the capacity control literature, the number of decision periods is not fixed, but may also depend on these external factors. We presented assumptions that ensure that the expected revenue maximizing policy can be described by protection levels. This means that given the current realization of demand and the external factor, the optimal action is to protect a certain number of seats for future requests and to sell as many remaining seats as possible.

This structure of an expected revenue maximizing policy is well-known in the dynamic and the static model, the two textbook models of single-resource capacity control. Indeed, we were able to show that these two models are special cases of the capacity control model in a random environment. But due to the more restrictive assumptions, monotonicity in the remaining time until departure and booking class could also be shown. We presented structures that are well-known in the literature and proved additional properties.

Traditionally, the aim of capacity control is to maximize expected revenue. We recapitulated the dynamic and the static model from the perspective of a risk-averse, expected utility maximizing decision-maker.

The standard approach to incorporating the notion of utility into Markov decision processes is to assume an additive time-separable utility function. Using this approach, we find that protection levels are suitable controls for the dynamic model but not for the static model. In the dynamic model, the protection levels can be seen to be monotone in time and booking class. In addition, they are monotone in the degree of risk-aversion. In the static model, optimal controls are still monotone in the remaining capacity, yet they are not of protection level type. In general, the controls need not be monotone in booking class or degree of risk-aversion.

These findings on an expected utility maximizing policy are counterintuitive at first glance, but we argued that they stem from the rather unrealistic preferences induced by an additive time-separable utility function. We suggested an atemporal utility function as a more realistic alternative. In general, optimal controls of an expected atemporal utility maximizing policy need not even be monotone in the remaining capacity. In the case of an exponential utility function with a positive coefficient of absolute risk-aversion, however, the expected utility maximizing policy is of protection level type and monotone in the booking class. In the dynamic model, it is also monotone in the remaining time. Furthermore, we were able to show that if there is a policy that is optimal for all sufficiently small values of γ, this policy is the optimal policy for a risk-neutral decision-maker.

Due to the high popularity of the EMSR heuristics for the static model, we suggested straightforward extensions to account for constant absolute risk-aversion and compared them to Weatherford's EMSU heuristic (Weatherford, 2004) in a small simulation study.

Finally, we showed that even in the more general capacity control model under a general discrete choice model of consumer behavior, the structural results known from the expected revenue maximizing setting carry over to the case of a decision-maker maximizing expected atemporal utility with an exponential utility function.

9.2 Directions for Future Work

Several directions for future work are straightforward; we highlight four of them in the following.

First, we did not show that protection levels of a γ-optimal policy are monotone in the coefficient of absolute risk-aversion γ for both the dynamic and the static model. Yet all of our numerical testing supports this hypothesis. For the dynamic model with only one item to sell, the monotonicity follows directly from the definition of increasing absolute risk-aversion. For the static model with only two booking classes, the result is shown in Eeckhoudt et al. (1995).

A second direction for future research might be a closer investigation of the dynamic and the static model for the maximization of expected atemporal utility with a general utility function on total revenue. By means of examples, we showed that structural properties of an expected utility maximizing policy depend heavily on the shape of the utility function. Judging from the relevance of these findings, it would be interesting to find necessary conditions that guarantee the existence of an atmp-optimal policy that is increasing in capacity and time.

Third, factors other than revenue might have an impact on the decision-maker's preferences. Lost customer goodwill, e.g. due to high spill rates, might lower demand for future flights (see e.g. Lindenmeier and Tscheulin, 2005, and Wirtz et al., 2003). Disappointed investors, e.g. due to a reduction of market share (determined by load factors) or volatile revenues earned, might increase the costs of outside capital in the long run or even cause venture capitalists to leave the company. In certain business scenarios, the impact of these factors on future revenue could be crucial and should be considered in the capacity allocation process. A first step in this direction is made in Barz et al. (2006).

Last, and probably most important with respect to real applications, we mainly considered basic capacity control models. These models account neither for cancelations nor for no-shows, which are very common phenomena in practice. In addition, we only discussed the single-resource setting. The transfer to network problems presents another challenge for future research. Finally, the impact of risk-sensitivity on dynamic pricing strategies could be investigated.

A

List of Symbols

Symbol	Description	page
$\|v\|$	Supremum norm $\sup_{x\in\mathfrak{S}} \|v(s)\|$	17
1	Vector with entries 1	24
$\succeq$	Rational preference relation	30
$=_{st}$	Equality in distribution	149
a^+	$\max\{0, a\}$	18
a	Feasible action	19, 22
$A_n v(x_1)$	$\sum_{x_2\in\mathfrak{X}_2} \hat{p}_n(x_2)v(x_1, x_2)$	21
$\tilde{A}_n v(x_1, x_3)$	$\sum_{x_2\in\mathfrak{X}_2} \hat{p}_n(x_2)v(x_1, x_2, x_3)$	96
α	Discount factor	20, 22
$\mathfrak{A}$	Action space	19, 22
$\mathfrak{A}_n$	Action space at time n	19
$\mathfrak{A}(x)$	Set of feasible actions in state x	22
$\mathfrak{A}_n(x)$	Set of feasible actions in state x at time n	19
b	A function $\mathfrak{X} \times \mathfrak{A}_n \to \mathfrak{X}_1$	21
B	Binomially distributed random variable	50, 149
β	Real-valued parameter of a utility function	30, 32, 37
$\mathfrak{B}$	Set of all bounded functions on $\mathfrak{J}$	24
c	Remaining capacity	49
C	Fixed capacity of a resource with homogeneous units	4
$\mathfrak{C}$	Set of all remaining capacity values	49
d	Realization of D	47
$d_{\max}$	Maximum number of requests per period	47
D	Number of requested reservations (size of demand)	48
$\delta_{ii'}$	Indicator function	56
$\Delta g(x)$	$g(x) - g(x-1)$	17
$\mathfrak{D}$	Range of D	47
e	Realization of E	25, 48
E	Environmental state	25, 48

η	Joint distribution of D and I given Z	48
$\mathfrak{E}$	State space of the environment	25, 48
$\mathbb{E}[R]$	Expectation of R	31
$\mathbb{E}_\pi[R]$	Expectation of R with respect to P_π	20
$\mathbb{E}[R\|X]$	Expectation of R conditioned by X	20
f	Decision rule	20, 22
f^∞	Stationary policy $(f, f, \dots)$	22
f^*	Optimal decision rule	21, 25
$f^{\text{add}*}$	Add-optimal decision rule	37
$f^{\text{atmp}*}$	Atmp-optimal decision rule	39
$f^{\gamma *}$	γ-optimal decision rule	41
F	Efficient frontier	128
F^γ	γ-efficient frontier	133
$\mathfrak{F}$	Set of all decision rules	22
$\mathfrak{F}_n$	Set of all decision rules at time n	20
$\mathfrak{F}_\emptyset$	Set of all decision rules with $f_n(c) = 0$ for $c \leq 0$	136
$\mathfrak{F}_y$	Set of all decision rules of protection level type	113
$\mathfrak{F}^N$	$\mathfrak{F}_0 \times \cdots \times \mathfrak{F}_{N-1}$, if $\mathfrak{F} = \mathfrak{F}_0 = \cdots = \mathfrak{F}_{N-1}$	20
$\mathfrak{F}^\infty$	$\mathfrak{F} \times \mathfrak{F} \times \dots$	22
g	A real valued function	17
G_n^γ	$-V_n^\gamma$	100
γ	Coefficient of absolute risk-aversion	32, 34
$\Gamma g(x)$	$g(x)/g(x-1)$	17
h	A real valued function	52
$Hv(x)$	$\max_{a \in \mathfrak{A}(x)} \sum_{x' \in \mathfrak{J}} p(x, a, x') v(x')$	24
$\mathfrak{H}$	An arbitrary set	17
i	Requested product or booking/ fare class	4
$i_{\max}$	Number of distinct products or booking/ fare classes	4
I	Requested product or booking/ fare class	48
$\mathfrak{I}$	Set of all distinct products or booking/ fare classes	47, 125
j	An index	76
$\mathfrak{J}$	Essential state space	24
$\mathfrak{J}_0$	Absorbing set	23
$\tilde{\mathfrak{J}}$	Essential state space of the environmental process	26
$\tilde{\mathfrak{J}}_0$	Absorbing set of the environmental process	26
k	Number of variables important to the decision-maker	30
κ	Distribution of Z_{n+1} given D_n, I_n and Z_n	48
$\mathfrak{K}$	Constraint set	22
$\mathfrak{K}_n$	Constraint set at time n	19
ℓ	Popularity of the destination	57
$Lv(x,a)$	$r(x,a) + \sum_{x' \in X} p(x, a, x') v(x')$	23
$\lambda(\mathfrak{M})$	Total probability of purchase when offering $\mathfrak{M}$	127
λ_j	Total probability of purchase when offering $\mathfrak{M}_j$	128
$\mathfrak{L}$	Set of all simple lotteries	30

m	Outcome of a lottery	30
$m_{\max}$	Number of possible outcomes of a simple lottery	30
μ	Number of efficient sets	128
$\mathfrak{M}$	Set of products offered	125
$\mathfrak{M}_j$	Efficient set	128
n	Decision period/ time	19, 22
N	Planning horizon	19
ν	Probability measure	56
$\mathfrak{N}$	Set of products offered	127
$\mathbb{N}_0$	Set of all non-negative integers	17
$\mathbb{N}$	Set of all positive integers	17
$\vec{\iota}$	Realization of $\vec{\mathcal{O}}$	30
$\vec{\mathcal{O}}$	Outcome of a lottery	30
$\omega(\mathfrak{M})$	Expected one-stage revenue from offering $\mathfrak{M}$	127
ω_j	Expected one-stage revenue from offering $\mathfrak{M}_j$	128
$\omega^\gamma(\mathfrak{M})$	Expected one-stage utility from offering $\mathfrak{M}$	132
ω_j^γ	Expected one-stage utility from offering $\mathfrak{M}_j$	133
p	Transition law	19,22
$\hat{p}$	Distribution of a random disturbance	21, 63, 70
$\tilde{p}$	Transition law of the environmental process	25
$\vec{p}$	Vector of probabilities $\vec{p} = (p_1, \ldots, p_M) \in \mathfrak{L}$	30
$P(A)$	Probability for event A	20
P_π	Product measure	20, 22
$\tilde{P}$	Transition matrix of the environmental process	25
$\tilde{P}_{\tilde{\mathfrak{z}}}$	Substochastic matrix	26
$\tilde{P}^{\mathfrak{z}}$	Transition matrix of the external process	55
$\mathcal{P}(\Im)$	Power set of $\Im$	126
π	Policy	20, 22
π^*	Optimal policy	20
$\pi^{\mathrm{add}*}$	Add-optimal policy	37
$\pi^{\mathrm{atmp}*}$	Atmp-optimal policy	39
$\pi^{\gamma*}$	γ-optimal policy	41
Π	Set of policies	20, 22
ϕ	Distribution	56
ψ	Cost function	50
ψ^{P}	One-stage penalty cost	49
ψ^{T}	Terminal cost	49
ψ^{DB}	Costs of denied boardings	50
$q^e_{c,c'}$	Cancelation probability	49
$q^{e^0}_{\mathrm{NS}}$	No-show probability	50
r	One-stage reward function	20, 22
$\bar{r}$	Bound of the one-stage reward function	20,22
r^e_n	Expected one-stage reward function	126
R	Total reward	20

R_π	Total reward by applying policy π	20,22,39
ϱ_i	Price/Fare of product i	4
$\bar{\varrho}$	A value substantially larger than ϱ_1	64
$\bar{\varrho}^u$	A value larger than $\max_{n=1,\ldots,N} u_n(\varrho_1)$	83, 88
s	Realization of S	25
S	System state	26
$\sigma[R]$	Standard deviation of R	74
$\mathfrak{S}$	System state space	25
T	Time until the resource is perished	4
τ	Entrance time into the absorbing set $\mathfrak{Z}_0$	24
ϑ	Parameter of a distribution	56
$\mathfrak{T}$	Set of parameter values	56
u	Utility function	30
u_γ	Exponential utility function with parameter γ	33
$Uv(x)$	$\max_{a\in\mathfrak{A}(x)} Lv(x,a)$	23
v	A real valued function	21
V	Value function	21,25
V^{add}	Value function	37
V^{atmp}	Value function	39
V^γ	Value function	41
V_N	Terminal reward if time is counted forward	20
V_0	Terminal reward if time is counted backwards	35
V_π^N	Conditional expected total reward	20
V_π	Conditional expected total reward	22
V^{N*}	Maximum expected total reward	21
V^*	Maximum expected total reward	23
$V^{\mathrm{add}*}$	Maximum expected time-additive utility	37
$V^{\mathrm{atmp}*}$	Maximum expected atemporal utility	39
$V^{\gamma*}$	Maximum expected atemporal exponential utility	39
$\bar{V}$	Bound of the terminal reward	20
$\mathrm{Var}[R]$	Variance of a random variable R	34
w	Realization of W	31
w_{cer}	Certainty equivalent	31
W	Decision-maker's final wealth	31
$\mathfrak{W}$	Range of W	38
x	Realization of X	19, 22
X	State	20, 22
$\mathfrak{X}$	State space	19, 22
y	Protection level	52
z	Realization of Z	48
Z	External state	48
$\mathfrak{Z}$	State space of external process	48
$\mathfrak{Z}^\ell$	State space of popularity process	57
$\mathbb{Z}$	Set of all integers	17

B

An Extension of Stidham's Lemma

According to Lemma 1 in Stidham (1978), the following holds

Lemma B.1.
Let $g : \mathbb{Z} \to \mathbb{R}$ be a concave function and define $f\colon \mathbb{Z} \to \mathbb{R}$ as

$$f(c) = \max_{a=0,\ldots,m} \{ar + g(c-a)\} \ ,$$

with $r \geq 0$ and $m \in \mathbb{N}_0$. Then, f is concave.

A direct proof can be found in Stidham (1978) and Lautenbacher and Stidham (1999, Lemma 1).

Since Lemma B.1 is a special case of the following extension, the proof of Lemma B.2 also proves Lemma B.1.

Lemma B.2 (An extension of Lemma 1 in Stidham, 1978).
Let $g : \mathbb{Z} \to \mathbb{R}$ and $u : \mathbb{R} \to \mathbb{R}$ be concave functions. Define $f\colon \mathbb{Z} \to \mathbb{R}$ as

$$f(c) = \max_{a=0,\ldots,m} \{u(a \cdot r) + g(c-a)\} \ ,$$

with $r \geq 0$ and $m \in \mathbb{N}_0$. Then f is concave.

Proof. Since $u(a \cdot r)$ is evaluated at a finite number of points $a \cdot r$ only, we can write $\breve{u}(c-a) \cdot r$ instead of $u(a \cdot r)$ with concave $\breve{u} : \mathbb{Z} \to \mathbb{R}$.

Setting $t = c - a$ yields

$$f(c) = \max_{c-m \leq t \leq c} \{\breve{u}(t)r + g(t)\} \ .$$

Let

$$t^* = \arg\max \{\breve{u}(t)r + g(t)\} \ .$$

From the concavity of g and $\breve{u}$, it follows that

$$f(c) = \begin{cases} \breve{u}(c-m)r + g(c-m) & t^* \leq c-m, \\ \breve{u}(t^*)r + g(t^*) & c-m < t^* < c, \\ \breve{u}(c)r + g(c) & t^* \geq c \, . \end{cases}$$

Given that $g(c)-g(c-1) \geq g(c+1)-g(c)$ and $\check{u}(c)-\check{u}(c-1) \geq \check{u}(c+1)-\check{u}(c)$ hold due to the assumed concavity, we obtain for $c < t^*$

$$\begin{aligned} f(c) - f(c-1) &= \check{u}(c)r + g(c) - \check{u}(c-1)r - g(c-1) \\ &= [\check{u}(c) - \check{u}(c-1)]r + g(c) - g(c-1) \\ &\geq [\check{u}(c+1) - \check{u}(c)]r + g(c+1) - g(c) \\ &= f(c+1) - f(c) \; . \end{aligned}$$

For $t^* \leq c \leq t^* + m$, it follows from the definition of t^* that

$$f(c) = \check{u}(t^*)r + g(t^*) \geq f(c-1)$$

and

$$f(c) = \check{u}(t^*)r + g(t^*) \geq f(c+1) \; .$$

Therefore,

$$f(c+1) - f(c) \leq 0 \leq f(c) - f(c-1) \; .$$

Owing to the assumed concavity of g and $\check{u}$, this can be rearranged for $c > t^* + m$ to read

$$\begin{aligned} f(c) - f(c-1) &= \check{u}(c-m)r + g(c-m) - \check{u}(c-1-m)r - g(c-1-m) \\ &= [\check{u}(c-m) - \check{u}(c-1-m)]r + g(c-m) - g(c-1-m) \\ &\geq [\check{u}(c+1-m) - \check{u}(c-m)]r + g(c+1-m) - g(c-m) \\ &= f(c+1) - f(c) \; . \end{aligned}$$

Summing things up, $f(c) - f(c-1)$ is decreasing, i.e. $f(c)$ is concave.

C

Stochastic Concavity and the Binomial Distribution

We show that (A2) holds if the increase in capacity can be modeled by a binomial distribution with parameter $C - c$ and (environment dependent) cancelation probability q^e.

Lemma C.1. *Let* $g : \mathfrak{C} \to \mathbb{R}$ *be an increasing and concave function and* $B(n, q^e)$ *a binomial random variable with non-negative parameter* n *and* $0 \leq q^e \leq 1$. *Then, given* $C \geq 0$, *the function*

$$c \to \sum_{c'=c}^{C} q^e_{c,c'} g(c') = \mathbb{E}\left[g(c + B(C - c, q^e))\right]$$

is increasing and concave for all $e \in \tilde{\mathfrak{J}}$.

The proof is in the spirit of Example 4.1 and Proposition 3.7 in Shaked and Shanthikumar (1988). We write $\hat{X}_1 =_{st} \hat{X}_2$ if the random variable $\hat{X}_1$ equals $\hat{X}_2$ in distribution.

Proof. Choose c_1, c_2, c_3, and c_4 such that

$$c_1 \leq c_2 \leq c_3 \leq c_4 \tag{C.1}$$

and

$$c_1 + c_4 = c_2 + c_3 \; . \tag{C.2}$$

Now define $\hat{Y}_1$, $\hat{Y}_2$, and $\hat{Y}_3$ as independent random variables

$$\begin{aligned} \hat{Y}_1 &= B(C - c_4, q^e) \\ \hat{Y}_2 &= B(c_4 - c_3, q^e) \\ \hat{Y}_3 &= B(2c_3 - c_1 - c_4, q^e) \; . \end{aligned}$$

If we set

$$\begin{aligned}
\hat{X}_1 &= c_1 + \hat{Y}_1 + 2\hat{Y}_2 + \hat{Y}_3 \\
\hat{X}_2 &= c_1 + c_4 - c_3 + \hat{Y}_1 + \hat{Y}_2 + \hat{Y}_3 \\
\hat{X}_3 &= c_3 + \hat{Y}_1 + \hat{Y}_2 \\
\hat{X}_4 &= c_4 + \hat{Y}_1 \ ,
\end{aligned}$$

we can conclude that

(1) $\hat{X}_j =_{st} c_j + B(C - c_j, q^e)$. This is obvious for $j = 4$. For $j = 1, 2, 3$, it is a consequence of the convolution property of binomial random variables with same probability parameter q^e and (C.2).
(2) $\hat{X}_1 + \hat{X}_4 = \hat{X}_2 + \hat{X}_3$ almost surely because both sides of the equality can be reduced to $c_1 + c_4 + 2\hat{Y}_1 + 2\hat{Y}_2 + \hat{Y}_3$.
(3) $[\hat{X}_1, \hat{X}_2, \hat{X}_3] \leq \hat{X}_4$, i.e. $\hat{X}_1$, $\hat{X}_2$, and $\hat{X}_3$ are smaller than or equal to $\hat{X}_4$, almost surely. This follows from $\hat{Y}_1 \leq C - c_4$, $\hat{Y}_2 \leq c_4 - c_3$, and $\hat{Y}_3 \leq 2c_3 - c_1 - c_4$ and the definition of $\hat{X}_j$.

From properties (1), (2), and (3) and the concavity of g

$$g(\hat{X}_1) + g(\hat{X}_4) \leq g(\hat{X}_2) + g(\hat{X}_3)$$

holds almost surely. Taking expectations yields

$$\mathbb{E}\left[g(\hat{X}_1)\right] + \mathbb{E}\left[g(\hat{X}_4)\right] \leq \mathbb{E}\left[g(\hat{X}_2)\right] + \mathbb{E}\left[g(\hat{X}_3)\right] \ .$$

Now as $\hat{X}_j =_{st} c_j + B(C - c_j, q^e)$, it follows that

$$\begin{aligned}
&\mathbb{E}\left[g(c_1 + B(C - c_1, q^e))\right] + \mathbb{E}\left[g(c_4 + B(C - c_4, q^e))\right] \\
&\leq \mathbb{E}\left[g(c_2 + B(C - c_2, q^e))\right] + \mathbb{E}\left[g(c_3 + B(C - c_3, q^e))\right] \ .
\end{aligned}$$

Consequently, $\mathbb{E}\left[g(c + B(C - c, q^e))\right]$ is concave in c. It follows directly from (3) that $\mathbb{E}\left[g(c + B(C - c, q^e))\right]$ is increasing in c.

References

AGRAWAL, V. and SESHADRI, S. (2000) Impact of uncertainty and risk aversion on price and order quantity in the newsvendor problem. *Manufacturing & Service Operations Management* **2**, 410–423.

ALLAIS, M. (1953) Le comportement de l'homme rationnel devant le risque: critique des postulats et axiomes de lécole Américaine. *Econometrica* **21**, 503–546.

ANDERSON, C. K. and WILSON, J. G. (2003) Wait or buy? The strategic consumer: pricing and profit implications. *Journal of the Operational Research Society* **54**, 299–306.

ARROW, K. J. (1965) *Aspects of the theory of risk-bearing.* Yrjoe Jahnssonin Saatio, Helsinki. Reprinted in Arrow (1971).

ARROW, K. J. (1971) *Essays in the theory of risk-bearing.* North-Holland, Amsterdam.

AVILA-GODOY, G., BRAU, A., and FERNANDEZ-GAUCHERAND, E. (1997) Controlled Markov chains with discounted risk-sensitive criteria: Applications to machine replacement. *Proceedings of the 36th IEEE Conference on Decision Control* 1115–1120.

AVILA-GODOY, G. and FERNANDEZ-GAUCHERAND, E. (2001) Risk-sensitive inventory control problems. *Proceedings of the 40th IEEE Conference on Decision and Control* 4577.

AVIV, Y. (2005) A partially observed Markov decision process for dynamic pricing. *Management Science* **51**, 1400–1416.

BADINELLI, R. D. (2000) An optimal, dynamic policy for hotel yield management. *European Journal of Operational Research* **121**, 476–503.

BAKER, W., MARN, M., and ZAWADA, C. (2001) Price smarter on the net. *Harvard Business Review* **79**, 122–127.

BALL, M. O. and QUEYRANNE, M. (2006) *Toward robust revenue management: competitive analysis of online booking.* Robert H. Smith School of Business Research Paper Series, University of Maryland.

BAMBERG, G. and COENENBERG, A. G. (2006) *Betriebswirtschaftliche Entscheidungslehre.* Reihe Betriebswirtschaft. Vahlen, München, 13th edition.

BARON, D. P. (1973) Point estimation and risk preferences. *Journal of the American Statistical Association* **68**, 944–950.

BARZ, C. (2006) How does risk aversion effect optimal revenue management policies? - A simulation study. In MATTFELD, D. C. and SUHL, L. (eds.), *Informationssysteme in Transport und Verkehr*, volume 4, 161–172. Books on Demand GmbH, Norderstedt.

BARZ, C. and SCHÖN, C. (2006) *On risk-sensitive revenue management under a general model of consumer choice behaviour.* Report WiOR-671, Universtät Karlsruhe.

BARZ, C., UHRIG-HOMBURG, M., and WALDMANN, K.-H. (2006) Yield management: beyond expected revenue. In DREIER, T., STUDER, R., and WEINHARDT, C. (eds.), *Information Management and Market Engineering*, volume 4 of *Studies on E-Organisation and Market Engineering*, 71–83. Universitätsverlag, Karlsruhe.

BARZ, C. and WALDMANN, K.-H. (2006) An application of Markov decision processes to the seat inventory control problem. In MORLOCK, M., SCHWINDT, C., TRAUTMANN, N., and ZIMMERMANN, J. (eds.), *Perspectives on operations research — Essays in honor of Klaus Neumann*, 113–128. Deutscher Universitäts-Verlag, Wiesbaden.

BARZ, C. and WALDMANN, K.-H. (2007) Risk-sensitive capacity control in revenue management. *Mathematical Methods of Operations Research* (forthcoming).

BELOBABA, P. P. (1987a) Air travel demand and airline seat inventory management. Ph.D. thesis, Massachusetts Institute of Technology, Cambridge.

BELOBABA, P. P. (1987b) Airline yield management. *Transportation Science* **21**, 63–73.

BELOBABA, P. P. (1989) Application of a probabilistic decision model to airline seat inventory control. *Operations Research* **37**, 183–197.

BELOBABA, P. P. (1992) *Optimal vs. heuristic methods for nested seat allocation.* Presentation at ORSA/TIMS Joint National Meeting, Cambridge.

BELOBABA, P. P. and WEATHERFORD, L. R. (1996) Comparing decision rules that incorporate customer diversion in perishable asset revenue management situations. *Decision Analysis* **27**, 343–363.

BEN-YOSEF, E. (2005) *The evolution of the US airline industry: Theory, strategy and policy.* Springer, New York.

BERNOULLI, D. (1738) Specimen theoriae novae de mensura sortis. *Commentarii Academiae Scientiarum Imperialis Petropolitannae* **5**, 175–192, reprinted in Bernoulli (1954).

BERNOULLI, D. (1954) Exposition of a new theory on the measurement of risk. *Econometrica* **22**, 23–36.

BERTSIMAS, D. and TSITSIKLIS, J. N. (1997) *Introduction to linear optimization.* Athena Scientific, Belmont.

BHATIA, A. and PAREKH, S. (1973) *Optimal allocation of seats by fare.* Presentation at the AGIFORS Reservations Study Group, Trans World Airlines.

BITRAN, G. R. and CALDENTEY, R. (2003) An overview of pricing models for revenue management. *Manufacturing & Service Operations Management* **5**, 203–229.

BITRAN, G. R. and GILBERT, S. M. (1996) Managing hotel reservations with uncertain arrivals. *Operations Research* **44**, 35–49.

BIYALOGORSKY, E., CARMON, Z., FRUCHTER, G. E., and GERSTNER, E. (1999) Research note: overselling with opportunistic cancellations. *Marketing Science* **18**, 605–610.

BIYALOGORSKY, E. and GERSTNER, E. (2004) Contingent pricing to reduce price risks. *Marketing Science* **23**, 145–155.

BIYALOGORSKY, E., GERSTNER, E., WEISS, D., and XIE, J. (2005) The economics of service upgrades. *Journal of Service Research* **7**, 234–244.

BLACKORBY, C., PRIMONT, D., and RUSSELL, R. R. (1998) Separability: a survey. In BARBERÁ, S., HAMMOND, P. J., and SEIDL, C. (eds.), *Handbook of utility theory: principles*, volume 1, 49–92. Kluwer, Boston.

BODILY, S. E. and WEATHERFORD, L. R. (1995) Perishable-asset revenue management: generic and multiple-price yield management with diversion. *Omega, International Journal of Management Science* **23**, 173–185.

BOTIMER, T. C. and BELOBABA, P. P. (1999) Airline pricing and fare product differentiation: a new theoretical framework. *The Journal of the Operational Research Society* **50**, 1085–1097.

BOUAKIZ, M. and KEBIR, Y. (1995) Target-level criterion in Markov decision processes. *Journal of Optimization Theory and Applications* **86**, 1–15.

BOUAKIZ, M. and SOBEL, M. J. (1992) Inventory control with an exponential utility criterion. *Operations Research* **40**, 603–608.

BOYD, E. A. (2004) Dramatic changes in distribution will require renewed focus on pricing and revenue management models. *Journal of Revenue and Pricing Management* **3**, 100–103.

BOYD, E. A. and BILEGAN, I. (2003) Revenue management and e-commerce. *Management Science* **49**, 1363–1386.

BOYD, E. A. and KALLESEN, R. (2004) The science of revenue management when passengers purchase the lowest available fare. *Journal of Revenue and Pricing Management* **3**, 171–177.

BRUMELLE, S. L. and MCGILL, J. I. (1993) Airline seat allocation with multiple nested fare classes. *Operations Research* **41**, 127–137.

BRUMELLE, S. L., MCGILL, J. I., OUM, T. H., SAWAKI, K., and TRETHEWAY, M. W. (1990) Allocation of airline seats between stochastically dependent demands. *Transportation Science* **24**, 183–192.

BRUMELLE, S. L. and WALCZAK, D. (2003) Dynamic airline revenue management with multiple semi-Markov demand. *Operations Research* **51**, 137–148.

CALDENTEY, R. and WEIN, L. M. (2005) *Revenue management of a make-to-stock queue.* Working paper, Stern School of Business, New York University, New York.

CHAN, L. M. A., SHEN, Z. J. M., SIMCHI-LEVI, D., and SWANN, J. L. (2004) Coordination of pricing and inventory decisions – A survey and classification. In SIMCHI-LEVI, D., WU, S. D., and SHEN, Z.-J. (eds.), *Handbook of quantitative supply chain analysis: modeling in the e-business era*, volume 74 of *International series in operations research management science*, 335–392. Kluwer, Boston.

CHATWIN, R. E. (1998) Multiperiod airline overbooking with a single fare class. *Operations Research* **46**, 805–819.

CHATWIN, R. E. (2002) *On the equivalence of pricing and discount allocation problems in revenue management.* Working paper, Delphinity LLC.

CHATWIN, R. E. (2003) *Pricing versus discount allocation in revenue management.* Working paper, Delphinity LLC.

CHEN, X., SIM, M., SIMCHI-LEVI, D., and SUN, P. (2004) *Risk aversion in inventory management.* Innovation in Manufacturing Systems and Technology (IMST). Working paper, MIT.

COOPER, W. L. and GUPTA, D. (2006) Stochastic comparisons in airline revenue management. *Manufacturing & Service Operations Management* **8**, 221–234.

COOPER, W. L., HOMEM-DE MELLO, T., and KLEYWEGT, A. J. (2006) Models of the spiral-down effect in revenue management. *Operations Research* **54**, 968–987.

COPELAND, D. G. and MCKENNEY, J. L. (1988) Airline reservations systems: lessons from history. *MIS Quarterly* **12**, 353–370.

CORALUPPI, S. P. (1997) Optimal control of Markov decision processes for performance and robustness. Ph.D. thesis, University of Maryland.

CORALUPPI, S. P. and MARCUS, S. I. (1999) Risk-sensitive and minimax control of discrete-time, finite-state Markov decision processes. *Automatica* **35**, 301–309.

CORNER, J. L. and CORNER, P. D. (1985) Characteristics of decision analysis practice. *The Journal of Operational Research Society* **46**, 304–314.

CORSTEN, H. and GÖSSINGER, R. (2005) Kapazitätssteuerung im Revenue Management. *Zeitschrift für Betriebswirtschaft – Ergänzungsheft* **1**, 31–52.

CROSS, R. (1998) *Revenue management: hard-core tactics for profit-making and market domination.* Orion Business, London.

CURRY, R. E. (1990) Optimal airline seat allocation with fare classes nested by origins and destinations. *Transportation Science* **24**, 193–204.

DANTHINE, J.-P. and DONALDSON, J. B. (2002) *Intermediate financial theory.* Prentice Hall Finance Series. Prentice Hall, New York.

DAUDEL, S. and VIALLE, G. (1992) *Yield-Management – Erträge optimieren durch nachfrageorientierte Angebotssteuerung.* Campus-Verlag, Frankfurt/Main.

EDGEWORTH, F. (1888) The mathematical theory of banking. *Journal of Royal Statistical Society* **51**, 113–127.

EECKHOUDT, L., GOLLIER, C., and SCHLESINGER, H. (1995) The risk-averse (and prudent) newsboy. *Management Science* **41**, 786–794.

EGUCHI, T. and BELOBABA, P. P. (2004) Modelling and simulation of impact of revenue management on Japan's domestic market. *Journal of Revenue and Pricing Management* **3**, 119–142.

EISENFÜHR, F. and WEBER, M. (2003) *Rationales Entscheiden.* Springer-Lehrbuch. Springer, Berlin, 4th edition.

ELMAGHRABY, W. and KESKINOCAK, P. (2003) Dynamic pricing in the presence of inventory considerations: research overview, current practices, and future directions. *Management Science* **49**, 1287–1309.

EPSTEIN, L. G. and ZIN, S. E. (1989) Substitution, risk aversion, and the temporal behavior of consumption and asset returns: a theoretical framework. *Econometrica* **57**, 937–969.

FAINBERG, E. A. (1982) Controlled Markov processes with arbitrary numerical criteria. *Theory of Probability and its Applications* **27**, 486–503.

FASSNACHT, M. and HOMBURG, C. (1998) Preisdifferenzierung und Yield Management bei Dienstleistungsanbietern. In MEYER, A. (ed.), *Handbuch Dienstleistungs-Marketing*, volume 1, 866–879. Schäffer-Poeschel, Stuttgart.

FARIAS, V. F. and VAN ROY, B. (2006) *Dynamic pricing with a prior on market response.* Working paper, Stanford University.

FARLEY, T. (2003) Groups need revenue management too. *Journal of Revenue and Pricing Management* **2**, 153–157.

FENG, Y. and GALLEGO, G. (2005) *Evaluating last seat availability contracts.* Working paper, Chinese University of Hong Kong.

FENG, Y. and XIAO, B. (1999) Maximizing revenues of perishable assets with a risk factor. *Operations Research* **47**, 337–341.

FENG, Y. and XIAO, B. (2000) Optimal policies of yield management with multiple predetermined prices. *Operations Research* **48**, 332–343.

FENG, Y. and XIAO, B. (2006a) A continuous-time seat control model for single-leg flights with no-shows and optimal overbooking upper bound. *European Journal of Operational Research* **174**, 1298–1316.

FENG, Y. and XIAO, B. (2006b) Integration of pricing and capacity allocation for perishable products. *European Journal of Operational Research* **168**, 17–34.

FISHBURN, P. C. (1970) *Utility theory for decision making*, volume 18 of *Publications in operations research.* John Wiley & Sons, New York.

FRENCH, S. (1986) *Decision theory: an introduction to the mathematics of rationality.* Ellis Horwood Limited, Chichester.

GALLEGO, G., KOU, S. G., and PHILLIPS, R. (2004) *Revenue management of callable products.* Working paper, Columbia University, New York.

GALLEGO, G. and PHILLIPS, R. (2004) Revenue management of flexible products. *Manufacturing & Service Operations Management* **6**, 321–337.

GERCHAK, Y., PARLAR, M., and YEE, T. K. M. (1985) Optimal rationing policies and production quantities for products with serveral demand classes. *Canadian Control Applications and Methods* **2**, 161–176.

GOLLIER, C. (2001) *The economics of risk and time.* MIT Press, Cambridge.

HALL, J. R., LIPPMAN, S. A., and MCCALL, J. J. (1979) Expected utility maximizing job search. In STEVEN A. LIPPMAN, J. J. M. (ed.), *Studies in the Economics of Search*, volume 123 of *Contributions to economic Analysis.* North-Holland, Amsterdam.

HARRIS, F. H. D. and PEACOCK, P. (1995) "Hold my place, please" - Yield management improves capacity-allocation guesswork. *Marketing Management* **4**, 34–46.

HELM, W. E. and WALDMANN, K.-H. (1984) Optimal control of arrivals to multiserver queues in a random environment. *Journal of Applied Probability* **21**, 602–615.

HINDERER, K. (1970) *Foundations of non-stationary dynamic programming with discrete time parameter*, volume 33 of *Lecture Notes in Operations Research and Mathematical Systems.* Springer, Berlin.

HINDERER, K. and WALDMANN, K.-H. (2001) Cash management in a randomly varying environment. *European Journal of Operational Research* **130**, 468–485.

HINDERER, K. and WALDMANN, K.-H. (2003) The critical discount factor for finite Markovian decision processes with an absorbing set. *Mathematical Methods of Operations Research* **57**, 1–19.

HINDERER, K. and WALDMANN, K.-H. (2005) Algorithms for countable state Markov decision models with an absorbing set. *SIAM Journal on Control and Optimization* **43**, 2109–2131.

HORNER, P. (2000) The Sabre story – The making of OR magic at AMR. *OR/MS Today* **27**, 46–47.

HOWARD, R. A. (1988) Decision analysis: practice and promise. *Management Science* **34**, 679–695.

HOWARD, R. A. and MATHESON, J. E. (1972) Risk-sensitive Markov decision processes. *Management Science* **18**, 356–369.

IBARRA-SALAZAR, J. (2003) The newsboy model: changes in risk and price. *The Geneva Risk and Insurance Review* **30**, 99–109.

JACOBSON, D. H. (1973) Optimal stochastic linear systems with exponential performance criteria and their relation to deterministic differential games. *IEEE Transactions on Automatic Control* **18**, 124–131.

KEENEY, R. L. (1982) Decision analysis: an overview. *Operations Research* **30**, 803–838.

KEENEY, R. L. and RAIFFA, H. (1976) *Decisions with multiple objectives: preferences and value tradeoffs.* Wiley series in probability and mathematical statistics. John Wiley & Sons, New York.

KENNEDY, O. S., HARDAKER, J. B., and QUIGGIN, J. (1994) Incorporating risk aversion into dynamic programming models: comment. *American Journal of Agricultural Economics* **76**, 960–964.

KERR, A. (1999) Utility maximising stochastic dynamic programming: an overview. *Proceedings of the 34th ORSNZ Conference* 59–68.

KERR, A. L., READ, E. G., and KAYE, R. J. (1998) Reservoir management with risk aversion. *Proceedings of the 33rd ORSNZ Conference* 167–176.

KIMES, S. E. (1989) Yield management: A tool for capacity constrained service firms. *Journal of Operations Management* **8**, 348–363.

KIMMS, A. and KLEIN, R. (2005) Revenue Management im Branchenvergleich. *Zeitschrift für Betriebswirtschaft – Ergänzungsheft* **1**, 1–30.

KIMMS, A. and MÜLLER-BUNGART, M. (2004) Erlössteigernde Verkaufssteuerung durch Absatzgrenzen für Logistikdienste mit mehreren knappen Ressourcen. *Automatisierungstechnik* 304–312.

KIRKWOOD, C. W. (2004) Approximating risk aversion in decision analysis applications. *Decision Analysis* **1**, 55–72.

KLEIN, R. (2001) Revenue Management: Quantitative Methoden zur Erlösmaximierung in der Dienstleistungsproduktion. *Betriebswirtschaftliche Forschung und Praxis* **3**, 245–259.

KLEIN, S. and LOEBBECKE, C. (2003) Emerging pricing strategies on the web: lessons from the airline industry. *Electronic Markets* **13**, 46–58.

KLEYWEGT, A. J. and PAPASTAVROU, J. D. (1998) The dynamic and stochastic knapsack problem. *Operations Research* **46**, 17–35.

KLEYWEGT, A. J. and PAPASTAVROU, J. D. (2001) The dynamic and stochastic knapsack problem with random sized items. *Operations Research* **49**, 26–41.

KNIGHT, F. H. (1921) *Risk, uncertainty and profit*, volume 31 of *Hart, Schaffner, and Marx Prize Essays.* Houghton Mifflin, Boston, reprinted in Knight (1985).

KNIGHT, F. H. (1985) *Risk, uncertainty and profit.* University of Chicago Press, Chicago, midway reprint edition.

KREPS, D. M. (1977a) Decision problems with expected utility critera, I: upper and lower convergent utility. *Mathematics of Operations Research* **2**, 45–53.

KREPS, D. M. (1977b) Decision problems with expected utility critera, II: stationarity. *Mathematics of Operations Research* **2**, 266–274.

KREPS, D. M. (1978) Sequential decision problems with expected utility criteria. III: upper and lower transience. *SIAM Journal on Control and Optimization* **16**, 420–429.

KREPS, D. M. (1988) *Notes on the theory of choice.* Underground classics in economics. Westview Press, Boulder.

KREPS, D. M. and PORTEUS, E. L. (1978) Temporal resolution of uncertainty and dynamic choice theory. *Econometrica* **46**, 185–200.

KREPS, D. M. and PORTEUS, E. L. (1979) Dynamic choice theory and dynamic programming. *Econometrica* **47**, 91–100.

LANCASTER, J. (2003) The financial risk of airline revenue management. *Journal of Revenue and Pricing Management* **2**, 158–165.

LAU, H.-S. (1980) The newsboy problem under alternative optimization objectives. *The Journal of the Operational Research Society* **31**, 525–535.

LAUTENBACHER, C. J. and STIDHAM, S. J. (1999) The underlying Markov decision process in the single-leg airline yield management problem. *Transportation Science* **33**, 136–146.

LEE, T. C. and HERSH, M. (1993) A model for dynamic airline seat inventory control with multiple seat bookings. *Transportation Science* **27**, 252–265.

LEVIN, Y., MCGILL, J., and NEDIAK, M. (2005) *Risk in revenue management and dynamic pricing.* Working paper, School of Business, Queen's University, Kingston.

LI, M. Z. F. and OUM, T. H. (2002) A note on the single leg, multifare seat allocation problem. *Transportation Science* **36**, 349–353.

LIANG, Y. (1999) Solution to the continuous time dynamic yield management model. *Transportation Science* **33**, 117–123.

LIKHITRUANGSILP, V. and IOANNOU, P. G. (2004) Risk-sensitive Markov decision process for underground construction planning and estimating. *International Conference on Computing in Civil and Building* **10**, 2004.06.02–04.

LIM, A. E. B. and SHANTHIKUMAR, J. G. (2007) Relative entropy, exponential utility, and robust dynamic pricing. *Operations Research* **55**, 198–214.

LIN, K. Y. (2004) A sequential dynamic pricing model and its applications. *Naval Research Logistics* **51**, 501–521.

LINDENMEIER, J. and TSCHEULIN, D. K. (2005) Kundenzufriedenheitsrelevante Effekte der Überbuchung im Rahmen des Revenue-Managements. *Zeitschrift für Betriebswirtschaft – Ergänzungsheft* **1**, 101–123.

LIPPMAN, S. A. (1975) Applying a new device in the optimization of exponential queuing systems. *Operations Research* **23**, 687–710.

LIPPMAN, S. A. and MCCALL, J. J. (1981) The economics of uncertainty: selected topics and probabilistic methods. In ARROW, K. J. and INTRILIGATOR, M. D. (eds.), *Handbook of mathematical economics*, volume 1, 211–284. North-Holland, New York.

LITTLEWOOD, K. (1972) Forecasting and control of passenger bookings. *AGIFORS Symposium Proceedings* **12**, 95–117.

LIU, Q. and VAN RYZIN, G. (2005) *Strategic capacity rationing to induce early purchases.* Decision, risk & operations working papers series DRO-2005-02, Columbia University, New York.

LIU, Y. (2005) Decision-theoretic planning under risk-sensitive planning objectives. Ph.D. thesis, Georgia Institute of Technology.

LUKASCHEWITSCH, M. (2005) *Revenue optimal risk-based upgrading.* Presentation at Operations Research 2005, Bremen.

MACHINA, M. J. (1987) Choice under uncertainty: problems solved and unsolved. *The Journal of Economic Perspectives* **1**, 121–154.

MAGLARAS, C. and MEISSNER, J. (2006) Dynamic pricing strategies for multi-product revenue management problems. *Manufacturing & Service Operations Management* **8**, 136–148.

MARCUS, S. I., FERNANDEZ-GAUCHERAND, E., HERNÁNDEZ-HERNÁNDEZ, D., CORALUPPI, S. P., and FARD, P. (1997) Risk sensitive Markov decision processes. In BYRNES, C. I., DATTA, B. N., and MARTIN, C. F. (eds.), *Systems and control in the twenty-first century*, volume 22, 263–279. Birkhäuser, New York.

MARKOWITZ, H. (1959) *Portfolio selection: efficient diversification of investments.* Yale University Press, New Haven.

MAYER, M. (1976) Seat allocation, or a simple model of seat allocation via sophisticated ones. *AGIFORS Symposium Proceedings* **16**, 103–125.

MCGILL, J. (1989) Optimization and estimation problems in airline yield management. Ph.D. thesis, University of British Columbia, Vancouver, Canada.

MCGILL, J. I. and VAN RYZIN, G. (1999) Revenue management: research overview and prospects. *Transportation Science* **33**, 233–256.

MIHATSCH, O. and NEUNEIER, R. (2002) Risk-sensitive reinforcement learning. *Machine Learning* **49**, 267–290.

MITRA, D. and WANG, Q. (2003) Stochastic traffic engineering, with applications to network revenue management. *IEEE INFOCOM 2003 – The Conference on Computer Communications* **22**, 396–405.

MITRA, D. and WANG, Q. (2005) Stochastic traffic engineering for demand uncertainty and risk-aware network revenue management. *IEEE/ACM Transactions on Networking* **13**, 221–233.

MOSSIN, J. (1969) A note on uncertainty and preferences in a temporal context. *American Economic Review* **59**, 172–174.

MÜLLER, A. (2000) Expected utility maximization of optimal stopping problems. *European Journal of Operational Research* **122**, 101–114.

NETESSINE, S. and SHUMSKY, R. A. (2005) Revenue management games: horizontal and vertical competition. *Management Science* **51**, 813–831.

VON NEUMANN, J. and MORGENSTERN, O. (1947) *Theory of games and economic behavior.* Princeton University Press, Princeton.

VON NITZSCH, R. (2002) *Entscheidungslehre – wie Menschen entscheiden und wie sie entscheiden sollten.* Schäffer-Poeschel, Stuttgart.

OHTSUBO, Y. and TOYONAGA, K. (2002) Optimal policy for minimizing risk models in Markov decision processes. *Journal of Mathematical Analysis and Applications* **271**, 66–81.

PAK, K. and PIERSMA, N. (2002) Overview of OR techniques for airline revenue management. *Statistica Neerlandica* **56**, 479–495.

PAPASTAVROU, J. D., RAJAGOPALAN, S., and KLEYWEGT, A. J. (1996) The dynamic and stochastic knapsack problem with deadlines. *Management Science* **42**, 1706–1718.

PETRUZZI, N. C. and DADA, M. (1999) Pricing and the newsvendor problem: a review with extensions. *Operations Research* **47**, 183–194.

PFEIFER, P. E. (1989) The airline discount fare allocation problem. *Decision Sciences* **20**, 149–157.

PHILLIPS, R. L. (2005) *Pricing and revenue optimization.* Stanford Business Books, Stanford.

POLLAK, R. A. (1967) Additive von Neumann-Morgenstern utility functions. *Econometrica* **35**, 485–494.

PÖLT, S. (1999) *Back to the roots: new results on leg optimization.* Presentation to the AGIFORS Reservations and Yield Management Study Group, London.

PRATT, J. W. (1964) Risk aversion in the small and in the large. *Econometrica* **32**, 122–136.

PUTERMAN, M. L. (1994) *Markov decision processes: discrete stochastic dynamic programming.* Wiley series in probability and mathematical statistics. Wiley, New York.

RICHTER, H. (1982) The differential revenue method to determine optimal seat allotments by fare type. *AGIFORS Symposium Proceedings* **22**, 339–362.

ROBERTS, A. W. and VARBERG, D. E. (1973) *Convex functions.* Academic Press, New York.

ROBINSON, L. W. (1995) Optimal and approximate control policies for airline booking with sequential nonmonotonic fare classes. *Operations Research* **43**, 252–263.

ROTHBLUM, U. G. (1984) Multiplicative Markov decision chains. *Mathematics of Operations Research* **9**, 6–24.

ROTHSTEIN, M. (1985) OR and the airline overbooking problem. *Operations Research* **33**, 237–248.

RUST, J. (1996) Numerical dynamic programming in economics. In AMMAN, H. M., KENDRICK, D. A., and RUST, J. (eds.), *Handbook of computational Economics*, Handbooks in Economics, 619–729. Elsevier, Amsterdam.

VAN RYZIN, G. and MCGILL, J. I. (2000) Revenue management without forecasting or optimization: an adaptive algorithm for determining airline seat protection levels. *Management Science* **46**, 760–775.

SAVAGE, L. J. (1954) *The foundations of statistics.* John Wiley & Sons, New York.

SCHÄL, M. (1975) Conditions for optimality in dynamic programming and for the limit of n-stage optimal policies to be optimal. *Zeitschrift für Wahrscheinlichkeitstheorie und verwandte Gebiete* **32**, 179–196.

SCHÄL, M. (1981) Utility functions and optimal policies in sequential decision problems. In MOESCHLIN, O. and PALLASCHKE, D. (eds.), *Game theory and mathematical economics*, 357–365. North-Holland, Amsterdam.

SCHMIDT, U. (2004) Alternatives to expected utility: some formal theories. In BARBERÁ, S., HAMMOND, P. J., and SEIDL, C. (eds.), *Handbook of utility theory: extensions*, volume 2, 757–838. Kluwer, Boston.

SCHWEITZER, M. E. and CACHON, G. P. (2000) Decision bias in the newsvendor problem with a known demand distribution: experimental evidence. *Management Science* **46**, 404–420.

SELDEN, L. (1978) A new representation of preferences over 'certain x uncertain' consumption pairs: the 'ordinal certainty equivalent' hypothesis. *Econometrica* **46**, 1045–1060.

SEN, A. and ZHANG, A. X. (1999) The newsboy problem with multiple demand classes. *IIE Transactions* **31**, 431–444.

SFODERA, F. (2006) *The spread of yield management practices.* Physica-Verlag, Heidelberg.

SHAKED, M. and SHANTHIKUMAR, J. G. (1988) Stochastic convexity and its applications. *Advances in Applied Probability* **20**, 427–446.

VAN SLYKE, R. and YOUNG, Y. (2000) Finite horizon stochastic knapsacks with applications to yield management. *Operations Research* **48**, 155–172.

SMITH, B. C., GUNTHER, D. P., RAO, B. V., and RATLIFF, R. M. (2001) E-commerce and operations research in airline planning, marketing, and distribution. *Interfaces* **31**, 37–55.

SMITH, B. C., LEIMKUHLER, J. F., and DARROW, R. M. (1992) Yield management at American Airlines. *Interfaces* **22**, 8–31.

SMITH, J. E. (1998) Evaluating income streams: a decision analysis approach. *Management Science* **44**, 1690–1708.

SPENCE, M. and ZECKHAUSER, R. (1972) The effect of the timing of consumption decisions and the resolution of lotteries on the choice of lotteries. *Econometrica* **40**, 401–403.

STARMER, C. (2000) Developments in non-expected utility theory: the hunt for a descriptive theory of choice under risk. *Journal of Economic Literature* **38**, 332–382.

STIDHAM, SHALER, J. (1978) Socially and individually optimal control of arrivals to a GI/M/1 queue. *Management Science* **24**, 1598–1610.

SUBRAMANIAN, J., LAUTENBACHER, C. J., and STIDHAM, S. (1999) Airline yield management with overbooking, cancellations, and no-shows. *Transportation Science* **33**, 147–167.

TALLURI, K. and VAN RYZIN, G. (2004a) Revenue management under a general discrete choice model of consumer behavior. *Management Science* **50**, 15–33.

TALLURI, K. T. and VAN RYZIN, G. J. (2004b) *The theory and practice of revenue management.* Kluwer, Boston.

TITZE, B. and GRIESSHABER, R. (1983) Realistic passanger booking behavior and the simple low-fare/high-fare seat allotment model. *AGIFORS Symposium Proceedings* **23**, 197–223.

TSCHEULIN, D. K. and LINDENMEIER, J. (2003) Yield-Management – Ein State-of-the-Art. *Zeitschrift für Betriebswirtschaft* **6**, 629–662.

VARIAN, H. R. (1992) *Microeconomic analysis.* W. W. Norton & Company, New York, 3rd edition.

WALCZAK, D. (2001) Dynamic modelling approaches to airline revenue management. Ph.D. thesis, University of British Columbia, Vancouver, Canada.

WALDMANN, K.-H. (1981) *Über dynamische Optimierungsprobleme bei stochastisch variierenden Umwelteinflüssen.* Habilitationsschrift, Technische Hochschule Darmstadt.

WALDMANN, K.-H. (1983) Optimal replacement under additive damage in randomly varying environments. *Naval Research Logistics Quarterly* **30**, 377–386.

WALDMANN, K.-H. (1984) Inventory control in randomly varying environments. *SIAM Journal on Applied Mathematics* **44**, 657–666.

WALDMANN, K.-H. (2006) On Markov decision models with an absorbing set. In DELLA RICCIA, G., DUBOIS, D., KRUSE, R., and LENZ, H.-J. (eds.), *Decision Theory and Mulit-Agent Planning*, 145–163. Springer, New York.

WEATHERFORD, L. R. (1997) Using prices more realistically as decision variables in perishable-asset revenue management problems. *Journal of Combinatorial Optimization* **1**, 277–304.

WEATHERFORD, L. R. (1998) A tutorial on optimization in the context of perishable-asset revenue management problems for the airline industry. In YU, G. (ed.), *Operations Research in the airline industry*, volume 1, 68–100. Kluwer, Boston, 4th edition.

WEATHERFORD, L. R. (2003) *EMSR vs. EMSU: revenue or utility?* Presentation at the Agifors Yield Management Study Group, Honolulu, Hawaii.

WEATHERFORD, L. R. (2004) EMSR versus EMSU: revenue or utility? *Journal of Revenue and Pricing Management* **3**, 277–284.

WEATHERFORD, L. R. and BODILY, S. E. (1992) A taxonomy and research overview of perishable asset revenue management: yield management, overbooking, and pricing. *Operations Research* **40**, 831–844.

WEATHERFORD, L. R., BODILY, S. E., and PFEIFER, P. E. (1993) Modeling the customer arrival process and comparing decision rules in perishable asset revenue management situations. *Transportation Science* **27**, 239–251.

WEIL, P. (1990) Nonexpected utility in macroeconomics. *The Quarterly Journal of Economics* **105**, 29–42.

WEISS, R. M. and MEHROTRA, A. K. (2001) Online dynamic pricing: efficiency, equity and the future of e-commerce. *Virginia Journal of Law and Technology* **6**.

WHITE, D. J. (1987) Utility, probabilistic constraints, mean and variance of discounted rewards in Markov decision processes. *OR Spektrum* **9**, 13 – 22.

WHITE, D. J. (1993a) *Markov decision processes.* Wiley, Chichester.

WHITE, D. J. (1993b) Minimising a threshold probability in discounted Markov decision processes. *Journal of Mathematical Analysis and Applications* **173**, 634–646.

WHITTLE, P. (1990) *Risk-sensitive optimal control.* Wiley-Interscience series in systems and optimization. Wiley, Chichester.

WIRTZ, J., KIMES, S. E., THENG, J. H. P., and PATTERSON, P. (2003) Revenue management: resolving potential customer conflicts. *Journal of Revenue and Pricing Management* **2**, 216–226.

WOLLMER, R. D. (1992) An airline management model for a single leg route when lower fare classes book first. *Operations Research* **40**, 26–37.

WU, C. and LIN, Y. (1999) Minimizing risk models in Markov decision processes with policies depending on target values. *Journal of Mathematical Analysis and Applications* **231**, 47–67.

YEOMAN, I. and MCMAHON-BEATTIE, U. (2004) *Revenue management and pricing: case studies and applications.* Thomson Learning, London.

YU, S. (1998) Optimization models for the first arrival target distribution function in discrete time. *Journal of Mathematical Analysis and Applications* **225**, 193–223.

ZACHARIAS, T. P. (1993) Representation of preferences in dynamic optimization models under uncertainty. In TAYLOR, C. R. (ed.), *Applications of dynamic programming to agricultural decision problems*, 11–22. Westview Press, Boulder [u.a.].

ZEA, M. (2002) Is airline industry risk unmanageable? *Mercer on Transport and Logistics* **9**, 21–26.

ZEHLE, K.-O. (1991) Yield management – eine Methode zur Umsatzsteigerung für Unternehmen der Tourismusindustrie. In SEITZ, E. and WOLF, J. (eds.), *Tourismusmanagement und -marketing*, 483–504. Verlag Moderne Industrie, Landsberg/Lech.

ZHANG, D. and COOPER, W. L. (2005) Revenue management for parallel flights with customer-choice behavior. *Operations Research* **53**, 415–431.

ZHAO, W. and ZHENG, Y.-S. (2001) A dynamic model for airline seat allocation with passenger diversion and no-shows. *Transportation Science* **35**, 80–98.

Lecture Notes in Economics and Mathematical Systems

For information about Vols. 1–508
please contact your bookseller or Springer-Verlag

Vol. 509: D. Hornung, Investment, R&D, and Long-Run Growth. XVI, 194 pages. 2002.

Vol. 510: A. S. Tangian, Constructing and Applying Objective Functions. XII, 582 pages. 2002.

Vol. 511: M. Külpmann, Stock Market Overreaction and Fundamental Valuation. IX, 198 pages. 2002.

Vol. 512: W.-B. Zhang, An Economic Theory of Cities.XI, 220 pages. 2002.

Vol. 513: K. Marti, Stochastic Optimization Techniques. VIII, 364 pages. 2002.

Vol. 514: S. Wang, Y. Xia, Portfolio and Asset Pricing. XII, 200 pages. 2002.

Vol. 515: G. Heisig, Planning Stability in Material Requirements Planning System. XII, 264 pages. 2002.

Vol. 516: B. Schmid, Pricing Credit Linked Financial Instruments. X, 246 pages. 2002.

Vol. 517: H. I. Meinhardt, Cooperative Decision Making in Common Pool Situations. VIII, 205 pages. 2002.

Vol. 518: S. Napel, Bilateral Bargaining. VIII, 188 pages. 2002.

Vol. 519: A. Klose, G. Speranza, L. N. Van Wassenhove (Eds.), Quantitative Approaches to Distribution Logistics and Supply Chain Management. XIII, 421 pages. 2002.

Vol. 520: B. Glaser, Efficiency versus Sustainability in Dynamic Decision Making. IX, 252 pages. 2002.

Vol. 521: R. Cowan, N. Jonard (Eds.), Heterogenous Agents, Interactions and Economic Performance. XIV, 339 pages. 2003.

Vol. 522: C. Neff, Corporate Finance, Innovation, and Strategic Competition. IX, 218 pages. 2003.

Vol. 523: W.-B. Zhang, A Theory of Interregional Dynamics. XI, 231 pages. 2003.

Vol. 524: M. Frölich, Programme Evaluation and Treatment Choise. VIII, 191 pages. 2003.

Vol. 525: S. Spinler, Capacity Reservation for Capital-Intensive Technologies. XVI, 139 pages. 2003.

Vol. 526: C. F. Daganzo, A Theory of Supply Chains. VIII, 123 pages. 2003.

Vol. 527: C. E. Metz, Information Dissemination in Currency Crises. XI, 231 pages. 2003.

Vol. 528: R. Stolletz, Performance Analysis and Optimization of Inbound Call Centers. X, 219 pages. 2003.

Vol. 529: W. Krabs, S. W. Pickl, Analysis, Controllability and Optimization of Time-Discrete Systems and Dynamical Games. XII, 187 pages. 2003.

Vol. 530: R. Wapler, Unemployment, Market Structure and Growth. XXVII, 207 pages. 2003.

Vol. 531: M. Gallegati, A. Kirman, M. Marsili (Eds.), The Complex Dynamics of Economic Interaction. XV, 402 pages, 2004.

Vol. 532: K. Marti, Y. Ermoliev, G. Pflug (Eds.), Dynamic Stochastic Optimization. VIII, 336 pages. 2004.

Vol. 533: G. Dudek, Collaborative Planning in Supply Chains. X, 234 pages. 2004.

Vol. 534: M. Runkel, Environmental and Resource Policy for Consumer Durables. X, 197 pages. 2004.

Vol. 535: X. Gandibleux, M. Sevaux, K. Sörensen, V. T'kindt (Eds.), Metaheuristics for Multiobjective Optimisation. IX, 249 pages. 2004.

Vol. 536: R. Brüggemann, Model Reduction Methods for Vector Autoregressive Processes. X, 218 pages. 2004.

Vol. 537: A. Esser, Pricing in (In)Complete Markets. XI, 122 pages, 2004.

Vol. 538: S. Kokot, The Econometrics of Sequential Trade Models. XI, 193 pages. 2004.

Vol. 539: N. Hautsch, Modelling Irregularly Spaced Financial Data. XII, 291 pages. 2004.

Vol. 540: H. Kraft, Optimal Portfolios with Stochastic Interest Rates and Defaultable Assets. X, 173 pages. 2004.

Vol. 541: G.-y. Chen, X. Huang, X. Yang, Vector Optimization. X, 306 pages. 2005.

Vol. 542: J. Lingens, Union Wage Bargaining and Economic Growth. XIII, 199 pages. 2004.

Vol. 543: C. Benkert, Default Risk in Bond and Credit Derivatives Markets. IX, 135 pages. 2004.

Vol. 544: B. Fleischmann, A. Klose, Distribution Logistics. X, 284 pages. 2004.

Vol. 545: R. Hafner, Stochastic Implied Volatility. XI, 229 pages. 2004.

Vol. 546: D. Quadt, Lot-Sizing and Scheduling for Flexible Flow Lines. XVIII, 227 pages. 2004.

Vol. 547: M. Wildi, Signal Extraction. XI, 279 pages. 2005.

Vol. 548: D. Kuhn, Generalized Bounds for Convex Multistage Stochastic Programs. XI, 190 pages. 2005.

Vol. 549: G. N. Krieg, Kanban-Controlled Manufacturing Systems. IX, 236 pages. 2005.

Vol. 550: T. Lux, S. Reitz, E. Samanidou, Nonlinear Dynamics and Heterogeneous Interacting Agents. XIII, 327 pages. 2005.

Vol. 551: J. Leskow, M. Puchet Anyul, L. F. Punzo, New Tools of Economic Dynamics. XIX, 392 pages. 2005.

Vol. 552: C. Suerie, Time Continuity in Discrete Time Models. XVIII, 229 pages. 2005.

Vol. 553: B. Mönch, Strategic Trading in Illiquid Markets. XIII, 116 pages. 2005.

Vol. 554: R. Foellmi, Consumption Structure and Macroeconomics. IX, 152 pages. 2005.

Vol. 555: J. Wenzelburger, Learning in Economic Systems with Expectations Feedback (planned) 2005.

Vol. 556: R. Branzei, D. Dimitrov, S. Tijs, Models in Cooperative Game Theory. VIII, 135 pages. 2005.

Vol. 557: S. Barbaro, Equity and Efficiency Considerations of Public Higer Education. XII, 128 pages. 2005.

Vol. 558: M. Faliva, M. G. Zoia, Topics in Dynamic Model Analysis. X, 144 pages. 2005.

Vol. 559: M. Schulmerich, Real Options Valuation. XVI, 357 pages. 2005.

Vol. 560: A. von Schemde, Index and Stability in Bimatrix Games. X, 151 pages. 2005.

Vol. 561: H. Bobzin, Principles of Network Economics. XX, 390 pages. 2006.

Vol. 562: T. Langenberg, Standardization and Expectations. IX, 132 pages. 2006.

Vol. 563: A. Seeger (Ed.), Recent Advances in Optimization. XI, 455 pages. 2006.

Vol. 564: P. Mathieu, B. Beaufils, O. Brandouy (Eds.), Artificial Economics. XIII, 237 pages. 2005.

Vol. 565: W. Lemke, Term Structure Modeling and Estimation in a State Space Framework. IX, 224 pages. 2006.

Vol. 566: M. Genser, A Structural Framework for the Pricing of Corporate Securities. XIX, 176 pages. 2006.

Vol. 567: A. Namatame, T. Kaizouji, Y. Aruga (Eds.), The Complex Networks of Economic Interactions. XI, 343 pages. 2006.

Vol. 568: M. Caliendo, Microeconometric Evaluation of Labour Market Policies. XVII, 258 pages. 2006.

Vol. 569: L. Neubecker, Strategic Competition in Oligopolies with Fluctuating Demand. IX, 233 pages. 2006.

Vol. 570: J. Woo, The Political Economy of Fiscal Policy. X, 169 pages. 2006.

Vol. 571: T. Herwig, Market-Conform Valuation of Options. VIII, 104 pages. 2006.

Vol. 572: M. F. Jäkel, Pensionomics. XII, 316 pages. 2006

Vol. 573: J. Emami Namini, International Trade and Multinational Activity. X, 159 pages, 2006.

Vol. 574: R. Kleber, Dynamic Inventory Management in Reverse Logistics. XII, 181 pages, 2006.

Vol. 575: R. Hellermann, Capacity Options for Revenue Management. XV, 199 pages, 2006.

Vol. 576: J. Zajac, Economics Dynamics, Information and Equilibnum. X, 284 pages, 2006.

Vol. 577: K. Rudolph, Bargaining Power Effects in Financial Contracting. XVIII, 330 pages, 2006.

Vol. 578: J. Kühn, Optimal Risk-Return Trade-Offs of Commercial Banks. IX, 149 pages, 2006.

Vol. 579: D. Sondermann, Introduction to Stochastic Calculus for Finance. X, 136 pages, 2006.

Vol. 580: S. Seifert, Posted Price Offers in Internet Auction Markets. IX, 186 pages, 2006.

Vol. 581: K. Marti; Y. Ermoliev; M. Makowsk; G. Pflug (Eds.), Coping with Uncertainty. XIII, 330 pages, 2006.

Vol. 582: J. Andritzky, Sovereign Default Risks Valuation: Implications of Debt Crises and Bond Restructurings. VIII, 251 pages, 2006.

Vol. 583: I.V. Konnov, D.T. Luc, A.M. Rubinov† (Eds.), Generalized Convexity and Related Topics. IX, 469 pages, 2006.

Vol. 584: C. Bruun, Adances in Artificial Economics: The Economy as a Complex Dynamic System. XVI, 296 pages, 2006.

Vol. 585: R. Pope, J. Leitner, U. Leopold-Wildburger, The Knowledge Ahead Approach to Risk. XVI, 218 pages, 2007 (planned).

Vol. 586: B.Lebreton, Strategic Closed-Loop Supply Chain Management. X, 150 pages, 2007 (planned).

Vol. 587: P. N. Baecker, Real Options and Intellectual Property: Capital Budgeting Under Imperfect Patent Protection. X, 276 pages , 2007.

Vol. 588: D. Grundel, R. Murphey, P. Panos , O. Prokopyev (Eds.), Cooperative Systems: Control and Optimization. IX, 401 pages , 2007.

Vol. 589: M. Schwind, Dynamic Pricing and Automated Resource Allocation for Information Services: Reinforcement Learning and Combinatorial Auctions. XII, 293 pages , 2007.

Vol. 590: S. H. Oda, Developments on Experimental Economics: New Approaches to Solving Real-World Problems. XVI, 262 pages, 2007.

Vol. 591: M. Lehmann-Waffenschmidt, Economic Evolution and Equilibrium: Bridging the Gap. VIII, 272 pages, 2007.

Vol. 592: A. C.-L. Chian, Complex Systems Approach to Economic Dynamics. X, 95 pages, 2007.

Vol. 593: J. Rubart, The Employment Effects of Technological Change: Heterogenous Labor, Wage Inequality and Unemployment. XII, 209 pages, 2007

Vol. 594: R. Hübner, Strategic Supply Chain Management in Process Industries: An Application to Specialty Chemicals Production Network Design. XII, 243 pages, 2007

Vol. 595: H. Gimpel, Preferences in Negotiations: The Attachment Effect. XIV, 268 pages, 2007

Vol. 596: M. Müller-Bungart, Revenue Management with Flexible Products: Models and Methods for the Broadcasting Industry. XXI, 297 pages, 2007

Vol. 597: C. Barz, Risk-Averse Capacity Control in Revenue Management. XIV, 163 pages, 2007

Vol. 598: A. Ule, Partner Choice and Cooperation in Networks: Theory and Experimental Evidence. Approx. 200 pages, 2007

Printing: Krips bv, Meppel
Binding: Stürtz, Würzburg